A SHORT BRIEF OF EXCAVATION

RAMCHANDRA SHRIVASTAV

DEDICATION

This book of mine is dedicated to all those people who believed in me and inspired me to continue writing. This book of mine is also dedicated to all those who took part in improving my writing by criticizing me. Expressing my gratitude and thanks to all of them many times, I dedicate this book to them and hope that my writing work will stand up to their standards.

PREFACE

Hello readers,

In this book I have tried to cover complete basic information of excavation work. I have tried to make it as simple as possible.

Excavation refers to the process of removing earth, rock, or other materials from a site in order to create a hole, trench, or cavity. It is a fundamental step in various construction, engineering, and archaeological activities. Excavation is a crucial initial step in many construction and engineering projects, and it requires careful planning, adherence to safety guidelines, and consideration of environmental and geological factors to ensure a successful and safe outcome

Excavation serves various purposes across different industries and applications. The specific purpose of excavation depends on the project or activity at hand. The purpose of excavation can vary significantly depending on the specific project's goals and requirements. It plays a vital role in construction, infrastructure development, resource extraction, environmental management, and scientific research. Proper planning and execution of excavation are essential to ensure the success and safety of these projects.

A SHORT BRIEF OF EXCAVATION
BY RAMCHANDRA SHRIVASTAV

TABLE OF CONTENTS

1

EXCAVATION

Excavation refers to the process of removing earth, rock, or other materials from a site in order to create a hole, trench, or cavity. It is a fundamental step in various construction, engineering, and archaeological activities. Here are some key aspects of excavation:

1.   Purpose: Excavation can serve various purposes, such as laying foundations for buildings, creating basements, digging trenches for utility lines (water, sewer, gas, etc.), mining minerals, extracting archaeological artifacts, or conducting geological studies.

2.   Methods: Excavation methods depend on the specific project and the type of material being removed. Common techniques include using heavy machinery like excavators, bulldozers, backhoes, or manual labor with shovels and pickaxes.

3.   Safety: Excavation can be hazardous, with the potential for cave-ins, equipment accidents, or encounters with underground utilities. Therefore, safety precautions, such as shoring, sloping, or trench boxes, are crucial to protect workers and ensure stability.

4.   Soil Analysis: Before excavation begins, it's essential to analyze the soil and geological conditions to determine the appropriate excavation methods and any necessary support structures.

5.   Environmental Considerations: Excavation can have environmental impacts, such as disturbing ecosystems or potentially contaminating soil or water. Proper planning and environmental assessments are necessary to mitigate these effects.

6.   Archaeological Excavation: In archaeology, excavation involves carefully removing layers of soil and sediment to uncover artifacts and archaeological remains. This process helps researchers study and preserve the past.

7.   Trench Excavation: Trenches are narrow, deep excavations typically used for utilities or foundations. Safety measures like sloping or shoring are critical in trench excavation to prevent collapse.

8.   Mass Excavation: Mass excavation involves removing large quantities of soil or rock, often to create foundations or prepare sites for construction.

9.   Grading: After excavation, grading may be necessary to level the ground and ensure proper drainage.

10.   Backfilling: Once the excavation's purpose is achieved, the hole or trench is typically backfilled

with soil or other suitable materials to restore the site's integrity.

Excavation is a crucial initial step in many construction and engineering projects, and it requires careful planning, adherence to safety guidelines, and consideration of environmental and geological factors to ensure a successful and safe outcome.

2

PURPOSE OF EXCAVATION

Excavation serves various purposes across different industries and applications. The specific purpose of excavation depends on the project or activity at hand. Here are some common purposes of excavation:

1.   Construction Foundations: Excavation is often performed to create the necessary depth and space for building foundations. This includes residential, commercial, and industrial structures.

2.   Basement Construction: Excavation may be undertaken to create underground basements or lower levels in buildings.

3.   Utility Installation: Excavation is essential for laying underground utility lines, such as water pipes, sewer lines, electrical conduits, and gas pipelines.

4.   Roads and Highways: Excavation is a critical part of road and highway construction to create the roadbed, cut through hills, and build bridges and tunnels.

5.   Mining: In the mining industry, excavation is the primary process for extracting valuable minerals, ores, and resources from the Earth.

6. **Landscaping:** Excavation is used in landscaping to shape the terrain, create ponds, and build features like retaining walls or terraces.

7. **Environmental Remediation:** Excavation can be employed to remove contaminated soil or materials from polluted sites as part of environmental cleanup efforts.

8. **Dredging:** Excavation in water bodies, known as dredging, is used to deepen channels, remove sediment, or extract minerals and materials from the seabed or riverbed.

9. **Archaeological Exploration:** Archaeologists use excavation to uncover and study historical artifacts and structures buried beneath the ground's surface.

10. **Geological Investigations:** Excavation may be performed for geological research and sampling purposes, helping scientists study Earth's subsurface layers.

11. **Land Development:** Before land development projects, excavation is conducted to prepare the site for residential, commercial, or recreational use.

12. **Trenching:** Excavation in the form of trenches is used for a wide range of purposes, including burying utility lines, installing drainage systems, or conducting soil analysis.

13.   **Erosion Control:** Excavation can help create terraces, berms, or other structures to prevent soil erosion and manage water runoff.

14.   **Demolition:** Excavation may be part of the demolition process, where structures are dismantled and removed from the site.

15.   **Pond or Reservoir Construction:** Excavation is necessary to create ponds or reservoirs for water storage, irrigation, or recreational purposes.

The purpose of excavation can vary significantly depending on the specific project's goals and requirements. It plays a vital role in construction, infrastructure development, resource extraction, environmental management, and scientific research. Proper planning and execution of excavation are essential to ensure the success and safety of these projects.

3

METHOD OF EXCAVATION

Excavation methods vary depending on the specific project, the type of material being excavated, and the equipment available. Here are some common methods of excavation:

1.   Mechanical Excavation:

•   Excavators: These versatile machines are equipped with buckets or attachments that can dig and remove soil, rocks, or other materials. They are commonly used in various types of excavation projects.

•   Backhoes: Backhoes combine a digging bucket on the back and a loader bucket on the front, making them useful for both excavation and material handling.

•   Bulldozers: Bulldozers are primarily used for pushing and clearing debris, but they can also perform shallow excavation by pushing materials to one side.

•   Dragline Excavators: These machines use a large bucket attached to a cable for digging and are often used in deep excavation projects or mining.

• **Trenchers:** Trenchers are specialized machines for digging long, narrow trenches, commonly used for installing utilities like pipes and cables.

• **Hydraulic Shovels:** These powerful machines use hydraulic systems to excavate and are often employed in large-scale mining and excavation projects.

2. **Manual Excavation:**

• **Shoveling:** Manual laborers use shovels to dig and remove soil, sand, or debris from the excavation site. This method is labor-intensive and typically used for smaller-scale projects.

• **Pickaxes and Mattocks:** These hand tools are useful for breaking up hard or compacted soil or rocks.

• **Manual Trenching:** In shallow trenching projects, manual labor with shovels and picks may be employed.

3. **Blasting:** Explosives are used to break up hard rock or other materials. Blasting is common in mining and large-scale excavation projects where mechanical methods are less effective.

4. **Tunneling:** Tunnel excavation involves creating underground passageways. Tunnel boring

machines (TBMs) are specialized equipment used for tunnel excavation.

5. Dredging: In underwater excavation, dredgers remove sediment, sand, or other materials from rivers, lakes, and seabeds using specialized equipment.

6. Hydro Excavation: This method uses pressurized water and a vacuum system to excavate soil and debris. It is often used when precision is required or when avoiding damage to existing utilities is crucial.

7. Pneumatic Excavation: Compressed air is used to loosen and remove soil or materials in areas where traditional digging methods might be too destructive or impractical.

8. Archaeological Excavation: Archaeologists use trowels, brushes, and other hand tools to carefully excavate historical or archaeological sites, preserving artifacts and context.

9. Rock Excavation: For projects involving solid rock, specialized rock-breaking equipment like hydraulic hammers or rock saws may be used.

The choice of excavation method depends on factors such as the project's scale, the type of material to be excavated, the desired precision, safety considerations, and environmental concerns. Proper planning, safety measures, and equipment selection are critical to the success and safety of excavation projects.

4

SAFETY MEASURES IN EXCAVATION

Safety is a paramount concern in excavation projects, as they can involve significant risks to both workers and the surrounding environment. Proper safety measures and protocols must be in place to prevent accidents and ensure the well-being of everyone involved. Here are some key safety considerations for excavation:

1.   Site Inspection and Assessment: Before starting any excavation work, the site should be thoroughly inspected to identify potential hazards, including underground utilities, unstable soil, nearby structures, and environmental concerns.

2.   Excavation Design and Planning: Adequate planning is essential, including determining the depth and dimensions of the excavation, the type of equipment to be used, and the necessary support structures (shoring, sloping, or trench boxes). Engineers or safety experts should be consulted when planning complex or deep excavations.

3.   Utility Locating: Underground utilities such as water pipes, gas lines, electrical cables, and sewer systems must be accurately located and marked

before digging begins to prevent accidental damage or injury.

4.   Protective Systems:

•   **Sloping:** The angle at which the excavation walls are sloped depends on the type of soil and the depth of the trench. Sloping helps prevent cave-ins.

•   **Shoring:** Support structures, such as steel or timber shoring, can be used to reinforce the excavation walls and prevent collapse.

•   **Trench Boxes:** These protective systems provide a secure enclosure for workers inside trenches and offer protection against cave-ins.

5.   Access and Egress: Safe entry and exit points should be provided for workers inside excavations, and ladders or ramps should be used when trenches are deeper than 4 feet (1.2 meters).

6.   Protective Equipment: Workers should wear appropriate personal protective equipment (PPE), including hard hats, high-visibility clothing, steel-toed boots, and safety goggles.

7.   Training and Education: All personnel involved in excavation work should receive proper training on excavation safety, including recognizing hazards, using equipment safely, and emergency procedures.

8. **Daily Inspections:** Regular inspections of the excavation site, equipment, and protective systems should be conducted to ensure they are in good working condition.

9. **Monitoring:** Soil conditions should be monitored continually to detect any signs of instability or movement. Instruments such as inclinometers can be used for this purpose.

10. **Emergency Response:** An emergency response plan should be in place, detailing procedures for evacuating workers, responding to injuries, and addressing potential hazards like gas leaks.

11. **Communication:** Effective communication among workers and with supervisors is crucial. Signals and clear instructions should be used to coordinate activities safely.

12. **Weather Conditions:** Weather can affect excavation safety. Rain can make soil unstable, and extreme heat or cold can pose risks to workers. Monitoring and adjustments to work schedules may be necessary.

13. **Environmental Protection:** Measures should be taken to prevent environmental damage, such as controlling erosion and runoff from the excavation site.

14. **Permits and Regulations: Ensure compliance with local, state, and federal regulations governing excavation work, and obtain any necessary permits.**

Safety should be an integral part of every excavation project, from the planning stage through to completion. Regular safety training, ongoing risk assessment, and a commitment to best practices are essential for preventing accidents and ensuring the safety of everyone involved.

5

SOIL ANALYSIS FOR EXCAVATION

Soil analysis is a critical step in the planning and execution of excavation projects. It helps engineers, geologists, and construction professionals understand the composition, properties, and stability of the soil at the excavation site. This information is essential for ensuring the safety and success of the project. Here are the key aspects of soil analysis for excavation:

1.   Sampling: Soil samples should be collected from various depths and locations within the excavation area. These samples can be obtained using soil borings, test pits, or auger drilling, depending on the project's requirements.

2.   Laboratory Testing: Soil samples are typically sent to a laboratory for comprehensive testing. The specific tests performed may include:

•   Particle Size Analysis: Determines the proportions of sand, silt, and clay in the soil, which affects its properties like permeability and cohesion.

•   Moisture Content Analysis: Measures the amount of water present in the soil, which influences its density and compaction characteristics.

- **Atterberg Limits: Identifies the plastic and liquid limits of the soil, helping to classify it as clay, silt, or sand and assess its behavior when wet or dry.**

- **Compaction Testing: Evaluates how well the soil can be compacted to achieve the desired density.**

- **Shear Strength Testing: Assesses the soil's ability to resist deformation and shear forces, which is crucial for stability analysis.**

- **Permeability Testing: Determines the soil's ability to allow the flow of water through it.**

- **Consolidation Testing: Measures how much the soil compresses under load over time, which is important for settling calculations.**

- **California Bearing Ratio (CBR) Test: Evaluates the soil's strength and its suitability for supporting roads and foundations.**

- **Swell Potential Testing: Determines whether the soil will expand significantly when wet, which is critical for foundations and pipelines.**

- **Chemical Analysis: Identifies the presence of contaminants or chemicals in the soil that could pose environmental concerns.**

3. **In-Situ Testing: In addition to laboratory tests, in-situ testing methods are used to assess soil**

properties at the excavation site. Common in-situ tests include:

• **Standard Penetration Test (SPT):** Measures the soil's resistance to penetration by a standardized drill rod, providing insights into soil density and bearing capacity.

• **Cone Penetration Test (CPT):** Measures soil resistance with a cone-shaped penetrometer, offering data on soil behavior, bearing capacity, and frictional properties.

• **Vane Shear Test:** Measures the in-situ shear strength of soft soils.

• **Plate Load Test:** Assesses the bearing capacity of the soil beneath a test plate, which can simulate the load from a structure.

4.   **Data Analysis:** Once the laboratory and in-situ test results are obtained, engineers and geologists analyze the data to determine soil classifications, properties, and behaviors. This analysis is used to design appropriate excavation support systems, foundations, and safety measures.

5.   **Report:** A comprehensive soil analysis report is generated, summarizing the findings, interpretations, and recommendations. This report guides the engineering and construction phases of the excavation project.

6.   **Ongoing Monitoring: In some cases, ongoing soil monitoring may be required during excavation to ensure that conditions remain within safe limits and that the project's design assumptions are validated.**

Soil analysis is a crucial aspect of excavation planning and execution because it informs decisions related to excavation methods, support systems, foundation design, and safety measures. It helps mitigate risks associated with soil instability, groundwater seepage, and other geological factors that can affect the project's success and safety.

6

ENVIRONMENTAL CONSIDERATIONS IN EXCAVATION

Excavation projects can have significant environmental impacts, and it's essential to consider and mitigate these effects to ensure responsible and sustainable construction practices. Here are some key environmental considerations in excavation:

1.   Environmental Impact Assessment (EIA): Conducting a comprehensive EIA before beginning any excavation project is essential, especially for larger projects. The EIA assesses potential environmental impacts, such as soil erosion, habitat disruption, and water quality issues, and proposes measures to mitigate these impacts.

2.   Erosion and Sediment Control: Excavation can lead to increased soil erosion if proper measures are not in place. Implement erosion control practices such as silt fences, sediment basins, and vegetative cover to prevent soil runoff into nearby water bodies.

3.   Stormwater Management: Develop a stormwater management plan to capture and treat rainwater runoff from the excavation site. This helps

prevent contamination of natural waterways with sediment, pollutants, or construction materials.

4.   Sediment and Water Quality Monitoring: Regularly monitor sediment and water quality in nearby streams, rivers, or ponds to ensure that excavation activities are not negatively affecting these ecosystems.

5.   Vegetation and Habitat Protection: If the excavation site contains sensitive vegetation or habitats, take measures to protect them. This may include establishing buffer zones, implementing fencing, or transplanting endangered plants to safe locations.

6.   Endangered Species Protection: Be aware of any endangered or protected species in the area and follow regulations to protect their habitats during excavation. This may include modifying project schedules to avoid nesting seasons or creating safe passageways for wildlife.

7.   Dust and Air Quality Control: Excavation can generate dust, which may contain particulate matter and other pollutants. Use dust control measures such as water sprays, dust suppressants, or covering materials to reduce airborne dust and maintain air quality.

8.   Noise and Vibration Control: Excavation activities can produce noise and vibrations that may impact nearby communities and wildlife. Implement

noise barriers and vibration monitoring to minimize disruptions.

9.   Waste Management: Properly manage and dispose of construction waste, including soil, rocks, and debris. Recycling and reusing materials can reduce the environmental impact.

10.   Hazardous Materials Handling: If the excavation site has potential hazardous materials or contaminants, such as buried tanks or polluted soil, take precautions to safely remove and dispose of these materials following environmental regulations.

11.   Restoration and Reclamation: Develop a plan for restoring the excavation site after the project is completed. This may involve regrading, reseeding, or replanting native vegetation to return the site to its natural state.

12.   Community Engagement: Engage with the local community to address their concerns, provide information about the project's environmental impacts, and solicit feedback. This can foster goodwill and help address potential issues early on.

13.   Regulatory Compliance: Ensure compliance with all applicable environmental laws, regulations, and permits. Failing to do so can result in legal and financial consequences.

14.   Sustainability Practices: Consider sustainable excavation practices, such as minimizing

soil disturbance, using environmentally friendly equipment, and exploring alternative construction methods that reduce the project's carbon footprint.

15. Environmental Education and Training: Provide training to project personnel on environmental best practices and the importance of minimizing environmental impacts.

By carefully considering and addressing these environmental considerations in excavation projects, construction professionals can help protect ecosystems, minimize disruption to local communities, and contribute to more sustainable and responsible construction practices.

7

ARCHAEOLOGICAL EXCAVATION

Archaeological excavation is a systematic and scientific process of uncovering, recording, and studying the material remains of past human cultures and civilizations. It is a crucial aspect of archaeology, which aims to reconstruct and understand the history and lifestyles of ancient societies by examining artifacts, structures, and other archaeological evidence. Here are the key components and steps involved in archaeological excavation:

1.   Site Selection: Archaeologists begin by selecting a site for excavation based on research, historical records, surveys, or remote sensing techniques. The site may be chosen for its historical significance, potential for discovery, or research questions it can help answer.

2.   Survey and Documentation: Before excavation begins, a site is thoroughly surveyed and documented. This involves mapping the site's boundaries, topography, and visible features, as well as conducting geophysical surveys to identify buried structures or artifacts.

3.   **Excavation Planning:** Archaeologists develop an excavation plan that outlines the goals, methodology, and resources required for the project. They also consider factors like the preservation of the site and the safety of the excavation team.

4.   **Grid System:** The site is divided into a grid system to facilitate precise recording and documentation of finds. Each grid square is carefully excavated and documented.

5.   **Stratigraphy:** Archaeologists pay close attention to the stratigraphy of the site, which involves studying the layers of sediment, soil, and debris. Each layer represents a different time period, and understanding their sequence is crucial for establishing a chronological framework.

6.   **Excavation Techniques:** Various tools and techniques are used to remove soil and uncover artifacts. These may include shovels, trowels, brushes, and even dental tools for delicate work. Hand excavation is often preferred to avoid damaging fragile artifacts.

7.   **Artifact Recovery:** When artifacts or features (such as walls or hearths) are encountered, they are carefully excavated, documented, and labeled with their precise location within the grid system.

8.   **Recording   and   Documentation:** Archaeologists keep meticulous records of their

findings, including photographs, drawings, notes, and measurements. This documentation is crucial for later analysis and interpretation.

9.   Artifact Conservation: Recovered artifacts are cleaned, stabilized, and preserved to prevent deterioration. In some cases, specialized conservators may be involved.

10.   Analysis: Once artifacts and other data are collected, they undergo various forms of analysis. This includes studying the artifacts' material composition, dating them, and interpreting their cultural and historical significance.

11.   Publication: The results of the excavation are often published in scholarly journals and reports. These publications contribute to the broader field of archaeology and are crucial for sharing knowledge with the academic community and the public.

12.   Ethical Considerations: Archaeologists must adhere to ethical principles, including respect for local communities, indigenous perspectives, and the preservation of cultural heritage. Collaboration with stakeholders and obtaining proper permits and permissions are essential.

13.   Site Preservation: After excavation, archaeologists work to preserve the site for future generations. This may involve backfilling, stabilizing structures, or creating interpretive displays.

A SHORT BRIEF OF EXCAVATION
BY RAMCHANDRA SHRIVASTAV

Archaeological excavation is a methodical and labor-intensive process that requires careful planning, scientific rigor, and attention to detail. It allows archaeologists to uncover and study the material remnants of past civilizations, contributing to our understanding of human history, culture, and evolution.

8

TRENCH EXCAVATION

Trench excavation is a specific type of excavation that involves the digging of long, narrow, and relatively deep channels or trenches in the ground. Trenches are commonly used in various construction, utility installation, and civil engineering projects. Here are the key aspects of trench excavation:

1.   Purpose: Trenches are excavated for various purposes, including:

•   Utility Installation: Trenches are often dug to install underground utilities such as water pipes, sewer lines, gas pipelines, electrical conduits, and fiber optic cables.

•   Foundation Work: Trenches may be dug for the footings and foundations of buildings, walls, or other structures.

•   Drainage: Trenches can be used to create drainage systems, diverting water away from buildings or areas prone to flooding.

•   Excavation for Archaeology: Archaeologists use trenches to explore and uncover historical or archaeological sites in a controlled and systematic manner.

• **Soil Sampling:** Trenches may be dug to collect soil samples for geotechnical analysis or environmental assessments.

2. **Trench Dimensions:** Trenches come in various sizes, and their dimensions depend on the specific project requirements. Trench width, depth, and length are determined by factors such as the type of utility being installed and local building codes.

3. **Safety Measures:**

• **Sloping:** To prevent trench collapses, trench walls are often sloped at a specific angle, which depends on the type of soil and trench depth. The slope provides stability.

• **Shoring:** In cases where trench sloping is not feasible due to space constraints or soil conditions, trench walls can be supported with shoring systems, such as steel or timber shoring, to prevent cave-ins.

• **Trench Boxes:** These protective structures are placed inside the trench to shield workers and prevent soil collapse. Trench boxes are especially important in deep trenches.

4. **Soil Analysis:** Before excavation, a soil analysis is conducted to assess the type and stability of the soil. This analysis helps determine the appropriate safety measures and excavation techniques.

5. **Dewatering: In some cases, groundwater may need to be pumped out of the trench to keep it dry during excavation.**

6. **Excavation Equipment: Various equipment can be used for trench excavation, including backhoes, excavators, trenchers, and smaller machines designed for narrow trench work.**

7. **Trench Inspection: Regular inspections are conducted throughout the excavation process to ensure the trench remains safe and stable. Inspections should take place before work begins each day and after any significant change in conditions.**

8. **Trench Backfilling: After the purpose of the trench is achieved (e.g., utility installation), it is backfilled with appropriate materials, compacted to the required density, and restored to its original condition.**

9. **Environmental Considerations: Trench excavation can have environmental impacts, especially in areas with sensitive ecosystems or habitats. Environmental best practices, such as erosion control measures, are often implemented.**

10. **Regulatory Compliance: Trench excavation must comply with local building codes, safety regulations, and permits. Authorities often inspect trench excavations to ensure compliance.**

A SHORT BRIEF OF EXCAVATION
BY RAMCHANDRA SHRIVASTAV

Trench excavation is a fundamental process in many construction and infrastructure projects. Proper planning, adherence to safety guidelines, and environmental considerations are essential to ensure the safety of workers, the protection of the environment, and the success of the project.

9

MASS EXCAVATION

Mass excavation refers to the process of removing large volumes of earth, rock, or other materials from a construction site or excavation area. It is typically performed to prepare the site for construction, infrastructure development, mining, or other purposes that require the removal of extensive quantities of material. Here are key aspects of mass excavation:

1.   Purpose: Mass excavation serves several purposes, including:

•   Creating building foundations by removing excess soil or rock.

•   Preparing sites for road, bridge, or dam construction.

•   Extracting valuable minerals, ores, or aggregates in mining operations.

•   Clearing land for large-scale infrastructure projects.

2.   Site Assessment: Before beginning mass excavation, a detailed site assessment is conducted. This assessment evaluates the soil and rock composition, geological conditions, groundwater

levels, and any potential environmental or safety concerns.

3.   Planning and Design: Engineers and project managers develop plans and designs for the mass excavation, taking into account the project's specific requirements, site conditions, and safety considerations.

4.   Excavation Methods:

•   Heavy Machinery: Mass excavation is typically carried out using heavy machinery such as excavators, bulldozers, loaders, haul trucks, and sometimes specialized equipment like scrapers or draglines.

•   Blasting: In mining or rock excavation, explosives may be used to break up hard rock before removal.

5.   Safety Measures: Safety is a paramount concern in mass excavation due to the large-scale nature of the work. Safety measures include:

•   Sloping or benching excavation walls to prevent collapses.

•   Implementing shoring systems or trench boxes in deep excavations.

•   Regular inspections and monitoring of excavation conditions.

• Training and protective equipment for workers.

6. Environmental Considerations: Mass excavation can have significant environmental impacts, such as habitat disruption, dust generation, and soil erosion. Environmental mitigation measures may include erosion control, dust suppression, and habitat restoration after excavation.

7. Material Handling: Excavated materials are often transported to designated disposal or storage areas. Efficient material handling processes are crucial to minimize delays and costs.

8. Environmental Regulations: Compliance with environmental regulations and permits is essential. Authorities may require environmental impact assessments and mitigation plans for large-scale excavation projects.

9. Site Reclamation: After the excavation is complete, the site may undergo reclamation efforts, which can involve reshaping the land, planting vegetation, and restoring the site to a condition suitable for its intended use.

10. Quality Control: Quality control measures ensure that the excavated material meets project specifications and is suitable for its intended purpose, such as fill material for construction or aggregates for concrete.

11. Monitoring and Reporting: Continuous monitoring of excavation progress and conditions is essential. Detailed reports are often generated to document excavation activities, materials removed, and any unexpected issues encountered.

12. Community Engagement: Depending on the project's size and impact, engaging with local communities and stakeholders to address concerns, provide information, and establish open communication channels is important.

Mass excavation is a significant undertaking that requires careful planning, adherence to safety and environmental regulations, and effective project management. It is often a crucial first step in large construction and infrastructure projects, setting the stage for further development.

10

GRADING AFTER EXCAVATION

Grading is the process of shaping and leveling the ground or site after excavation or construction work to ensure proper drainage, stability, and aesthetics. Grading is a crucial step in many construction and landscaping projects. Here are the key aspects of grading after excavation:

1. **Purpose:**

• **Surface Smoothness:** Grading ensures a smooth, even surface for construction, landscaping, or other planned activities.

• **Drainage Control:** Proper grading directs surface water away from buildings, structures, or areas where water accumulation could cause damage or flooding.

• **Erosion Prevention:** Grading helps control soil erosion by shaping the land to minimize water runoff and sediment transport.

• **Landscaping and Aesthetics:** Grading can be used to create desired landforms, slopes, and contours for landscaping, gardens, and outdoor spaces.

• **Accessibility:** Grading may be necessary to provide accessibility for vehicles, pedestrians, and those with disabilities.

2. **Site Assessment:** Before grading, assess the site's topography, soil composition, and drainage patterns. Understanding these factors is critical for effective grading design.

3. **Grading Plan:** Develop a grading plan that specifies the desired slopes, contours, and drainage patterns for the site. This plan should adhere to local regulations and engineering standards.

4. **Equipment and Materials:**

• **Heavy Machinery:** Grading is typically performed using heavy machinery such as bulldozers, graders, compactors, and skid-steer loaders.

• **Topsoil:** If necessary, topsoil may be brought in to provide a fertile layer for planting and landscaping.

5. **Cut and Fill:** Grading often involves a cut-and-fill process. Cut refers to excavating or removing soil from higher areas of the site, while fill involves adding soil to lower areas to achieve the desired grades.

6. **Slope Grading:** The slope of the graded surface depends on the project's requirements and

local regulations. For example, slopes around buildings may need to divert water away from structures, while landscaping features may have gentler slopes for aesthetics.

7.   Drainage Design: Proper drainage is crucial in grading. Grading should ensure that water flows away from buildings and structures and toward appropriate drainage systems, such as swales, catch basins, or stormwater management facilities.

8.   Compaction: Compaction of the graded soil is essential to ensure stability and prevent settling over time. Compaction equipment like rollers or compactors is used to achieve the desired soil density.

9.   Quality Control: Regular inspections and measurements are conducted during and after grading to ensure that the specified grades, slopes, and drainage patterns are achieved.

10.   Vegetation and Landscaping: If the project includes landscaping or planting, grading may be followed by soil preparation and planting of vegetation, grass, or trees to stabilize the soil and enhance the site's appearance.

11.   Erosion Control: Implement erosion control measures, such as silt fences, erosion mats, or hydroseeding, to prevent soil erosion during and after grading.

12.   Final Inspection and Documentation: After grading is completed, conduct a final inspection to ensure that the site meets the project's specifications. Document the work for future reference.

Grading is a critical step in site development, ensuring that the land is properly prepared for construction, landscaping, or other activities. It requires careful planning, precise execution, and adherence to local regulations and engineering standards to achieve the desired outcomes for both functionality and aesthetics.

11

COMPACTION AFTER EXCAVATION

Compaction after excavation is a crucial construction process that involves increasing the density of the soil or fill material in an excavated area. Proper compaction ensures that the soil is stable, has sufficient load-bearing capacity, and minimizes the risk of settling or subsidence. This is particularly important for construction projects where foundations, pavements, or structures will be built on the compacted soil. Here are the key aspects of compaction after excavation:

1.   Purpose:

•   Stability: Compaction improves the stability and load-bearing capacity of the soil, preventing settlement or settling of the ground beneath structures.

•   Density: It increases the density of the soil, reducing its permeability and making it less prone to water infiltration.

•   Uniformity: Compaction ensures uniform soil density, which is crucial for the even distribution of loads from buildings, roads, or other structures.

• **Erosion Control:** Compacted soil is less susceptible to erosion, which is important for preventing soil loss during heavy rainfall.

2. **Types of Compaction:**

• **Mechanical Compaction:** This method uses heavy machinery such as compactors, rollers, or vibratory plates to apply dynamic or static forces to the soil, reducing air voids and increasing soil density.

• **Pneumatic Compaction:** Pneumatic rollers use air pressure to compact the soil. They are effective for achieving compaction in loose or granular soils.

• **Vibrocompaction:** Vibratory probes or vibrators are inserted into the soil, which induces vibrations that settle the soil particles closer together.

• **Dynamic Compaction:** This involves dropping heavy weights from a height onto the soil surface to achieve compaction. It is often used for large-scale projects with loose or soft soils.

3. **Compaction Testing:** Compaction is not a one-size-fits-all process, and the level of compaction required depends on the project specifications and soil type. Compaction tests, such as the Proctor test or Modified Proctor test, are conducted to determine the optimum moisture content and

maximum dry density for the specific soil material. These tests help in developing a compaction plan.

4.   Moisture Content: Proper moisture content is crucial for effective compaction. Soil that is too dry may not compact well, while soil that is too wet can become overly compacted and less stable. The moisture content is adjusted to fall within the specified range.

5.   Compaction Equipment: The choice of compaction equipment depends on the type of soil and the project's requirements. Heavy vibratory rollers, sheepsfoot rollers, and plate compactors are commonly used for compaction.

6.   Layer-by-Layer Compaction: Soil is typically compacted in layers, with each layer being of a specified thickness. The compaction equipment makes multiple passes over each layer until the desired density is achieved.

7.   Quality Control: Compaction efforts are closely monitored throughout the process. Density tests and moisture content measurements are taken at various locations to ensure that the specified compaction requirements are met.

8.   Documentation: Detailed records of compaction efforts, including test results, moisture content adjustments, and equipment specifications, are maintained for quality assurance and project documentation.

A SHORT BRIEF OF EXCAVATION
BY RAMCHANDRA SHRIVASTAV

Proper compaction is essential for the long-term stability and performance of construction projects. It helps prevent issues such as settlement, uneven settling, and structural damage. Compaction should be carried out according to project specifications, and quality control measures should be rigorously followed to ensure that the desired level of compaction is achieved uniformly throughout the site.

12

BACKFILLING AFTER EXCAVATION

Backfilling is a construction process that involves filling an excavated area with suitable material to restore the ground to its intended grade, level, and stability. Backfilling is typically carried out after excavation for various purposes, including creating a foundation for structures, supporting underground utilities, and landscaping. Here are the key aspects of backfilling after excavation:

1.   Purpose:

•   Foundation Support: Backfilling is often done to provide a stable base for building foundations, footings, or retaining walls.

•   Utility Installation: After laying underground utilities like water pipes, sewer lines, or electrical conduits, backfilling is necessary to cover and protect them.

•   Erosion Control: Backfilling helps to prevent soil erosion by restoring the terrain's original contours.

• **Landscaping:** In landscaping projects, backfilling is used to create desired landforms, such as mounds or terraces.

• **Compaction:** Properly backfilled and compacted soil provides a stable surface for future construction or landscaping activities.

2. **Backfill Material:**

• The choice of backfill material depends on the project's requirements and the type of excavation. Common backfill materials include soil, sand, gravel, crushed stone, or engineered fill materials.

• Backfill materials should be free from debris, organic matter, or contaminants that could compromise the integrity of the fill or cause future settlement.

3. **Backfilling Process:**

• **Layer-by-Layer:** Backfilling is typically done in layers, with each layer compacted to a specified density before the next layer is added.

• **Moisture Control:** Proper moisture content in the backfill material is crucial for achieving adequate compaction. Excessively dry or wet material can lead to settlement issues.

• **Compaction:** Backfill material is compacted using heavy machinery, such as rollers or compactors, to ensure uniform density and stability.

• **Thick Lifts:** For some projects, thick lifts of backfill material may be used, with each lift compacted before adding the next.

4.   Utilities Protection: When backfilling around underground utilities, care should be taken to avoid damaging or displacing them. Utilities are often encased in protective materials or surrounded by a protective layer of sand or pea gravel.

5.   Slope Grading: The backfilled area is often graded to establish slopes that ensure proper surface drainage away from structures or other areas where water accumulation may cause issues.

6.   Compaction Testing: Compaction tests are conducted to verify that the backfilled material meets the specified compaction requirements. These tests help ensure that the soil is adequately compacted and stable.

7.   Inspection and Documentation: The backfilling process is closely inspected, and records are maintained to document the type and source of backfill material, compaction methods, moisture content adjustments, and test results.

8.   Settlement Allowance: It is common to anticipate some degree of settlement after

backfilling, especially with newly constructed foundations. Engineers may design for this anticipated settlement, and it can be monitored over time.

9.   Final Grading: After backfilling is complete, the final grading is carried out to achieve the desired surface contours and prepare the area for subsequent construction or landscaping activities.

Backfilling is a critical construction process that helps ensure the stability, safety, and functionality of various projects. Properly executed backfilling is essential to avoid future settlement issues, protect underground utilities, and create a suitable foundation for structures and landscapes.

13

MECHANICAL EXCAVATION

Mechanical excavation refers to the process of removing earth, rock, or other materials from a construction site or excavation area using mechanical equipment, such as excavators, bulldozers, backhoes, and trenchers. This method of excavation is widely used in construction, mining, road building, and various civil engineering projects due to its efficiency and precision. Here are the key aspects of mechanical excavation:

1. Equipment Selection: The choice of excavation equipment depends on factors such as the type of material to be excavated, the project's scale, site conditions, and specific requirements. Common types of equipment used for mechanical excavation include:

• Excavators: Versatile machines equipped with a bucket and arm for digging and loading materials.

• Bulldozers: Heavy-duty machines used for pushing, leveling, and clearing debris.

• **Backhoes:** Combines the functions of a loader and an excavator and is often used for digging and trenching.

• **Trenchers:** Specialized machines designed for digging trenches, commonly used in utility installation.

• **Wheel Loaders:** Used for loading and transporting materials within the construction site.

• **Skid-Steer Loaders:** Compact, maneuverable machines used for various tasks, including excavation, grading, and material handling.

2. **Excavation Techniques:** Different excavation techniques are employed depending on the equipment used and the project's requirements. These techniques may include digging, scraping, cutting, trenching, or grading.

3. **Safety Measures:** Safety is a paramount concern in mechanical excavation. Operators and workers must follow safety protocols, wear appropriate personal protective equipment (PPE), and be trained in equipment operation and safety practices.

4. **Site Preparation:** Before mechanical excavation begins, the site should be properly prepared, including clearing debris, marking utility lines, and ensuring proper access and egress routes for equipment.

5.   **Environmental Considerations:** Mechanical excavation can have environmental impacts, such as soil erosion, dust generation, and disruption to local ecosystems. Mitigation measures, such as erosion control, dust suppression, and habitat protection, may be necessary.

6.   **Material Handling:** Excavated materials are transported and placed as needed for the project. Efficient material handling processes help minimize delays and improve productivity.

7.   **Site Cleanup:** After mechanical excavation is complete, the site should be cleaned and cleared of debris, unused materials, and equipment to ensure safety and facilitate subsequent construction or development phases.

8.   **Regulatory Compliance:** Excavation projects must comply with local, state, and federal regulations and permits, especially when they involve environmental impact, utility installation, or land disturbance.

9.   **Monitoring and Inspection:** Regular monitoring and inspections are conducted to ensure the excavation work is proceeding according to the project's specifications and safety standards.

10.   **Quality Control:** Quality control measures may include checking the dimensions and depths of excavations, monitoring soil conditions, and

verifying that material compaction meets project requirements.

Mechanical excavation offers several advantages, including efficiency, precision, and the ability to handle a wide range of excavation tasks. However, it requires skilled operators and adherence to safety and environmental regulations to ensure successful and safe excavation operations.

14

MANUAL EXCAVATION

Manual excavation, also known as hand excavation, involves the removal of earth, soil, or other materials from a construction site or excavation area using hand tools and labor-intensive methods. This method of excavation is typically used for smaller-scale projects, archaeological digs, and in situations where mechanical equipment is impractical or unavailable. Here are the key aspects of manual excavation:

1.   Tools and Equipment: Manual excavation relies on basic hand tools, which may include:

•   Shovels and spades for digging and moving soil.

•   Picks and mattocks for breaking up compacted or rocky soil.

•   Trowels and hand shovels for precision work.

•   Buckets, wheelbarrows, or baskets for transporting excavated material.

•   Sieves and screens for sifting through soil to find artifacts or valuable items in archaeological excavations.

2.   Purpose: Manual excavation is often chosen for projects that require a high degree of precision or care, such as archaeological digs, small landscaping jobs, or when working in confined spaces where heavy machinery cannot be used.

3.   Labor Requirements: Manual excavation is labor-intensive and requires a team of workers to complete tasks efficiently. The number of laborers needed depends on the project's scope and the type of material being excavated.

4.   Precision and Control: Manual excavation allows for greater control and precision, making it suitable for delicate operations where the risk of damaging nearby structures or artifacts is high.

5.   Safety Measures: Safety is a significant concern in manual excavation. Workers must be trained in proper lifting techniques and be provided with personal protective equipment (PPE). Trenches and excavation sites must be properly shored or sloped to prevent cave-ins or collapses.

6.   Environmental Considerations: Manual excavation generally has fewer environmental impacts than mechanical excavation methods. However, precautions should still be taken to prevent soil erosion, dust generation, and habitat disruption.

7.   Material Handling: Manual excavation involves digging, lifting, and transporting soil and

materials using human power. Workers should be mindful of ergonomic considerations to prevent injuries.

8.   Excavation Techniques: Manual excavation techniques may include digging, scraping, grading, and leveling using hand tools. In archaeological excavations, trowels and brushes are used for delicate work around artifacts.

9.   Archaeological   Excavation: Manual excavation is commonly used in archaeological fieldwork. Archaeologists carefully excavate layers of soil to uncover artifacts and gain insights into historical or prehistoric cultures.

10.   Soil Analysis: Soil samples collected during manual excavation may be analyzed to understand the stratigraphy, composition, and archaeological context of the site.

11.   Site Cleanup: After manual excavation is complete, the site should be cleaned and cleared of debris, unused materials, and tools to ensure safety and facilitate subsequent phases of the project.

Manual excavation requires skilled labor, patience, and attention to detail. It is well-suited for projects where precision is essential or where mechanical equipment would be impractical or disruptive. However, it is generally slower and more labor-intensive than mechanical excavation methods, making it more suitable for smaller-scale tasks.

15

BLASTING

Blasting is a controlled explosion or detonation of explosives to break rock, concrete, or other hard materials in construction, mining, quarrying, and demolition applications. It is a technique used to fragment large masses of material into smaller, manageable pieces. Blasting is carried out under strict safety regulations and requires specialized knowledge and equipment. Here are the key aspects of blasting:

1.   Purpose:

•   Rock Excavation: Blasting is commonly used in mining and quarrying to break up rock formations for easier excavation and extraction of valuable minerals or aggregates.

•   Demolition: Blasting can be employed in controlled building demolitions to bring down structures safely and efficiently.

•   Construction: In construction, blasting may be used to excavate rock or create trenches for foundations in areas with hard, unyielding substrates.

• Road and Tunnel Construction: Blasting is often used to create tunnels and road cuts through rock formations.

• Dams and Reservoirs: Blasting can help clear bedrock and create foundations for large dams and reservoirs.

• Land Clearing: In some cases, blasting is used for land clearing or to break up large boulders that impede construction or development.

2. Safety:

• Safety is paramount in blasting operations. Strict safety protocols, regulations, and guidelines are followed to protect workers and the surrounding environment.

• Blast areas are typically restricted, and warning signals are used to ensure that no one is in the vicinity during detonation.

• Seismic monitoring and vibration control measures are implemented to prevent damage to nearby structures.

• Proper storage and handling of explosives are essential to minimize the risk of accidents.

3. Explosives:

• Various types of explosives are used in blasting, including dynamite, ammonium nitrate and fuel oil (ANFO), emulsion explosives, and blasting agents.

• The choice of explosive depends on factors such as the type of rock, desired fragmentation, and regulatory requirements.

4. Blasting Design:

• Blasting engineers or specialists carefully plan the blast, considering factors like the rock's hardness, the desired fragmentation size, and the proximity of structures or sensitive areas.

• Blast holes are drilled into the material, and explosive charges are strategically placed within these holes.

• The sequence and timing of detonation are critical for achieving the desired results while minimizing shock waves, vibrations, and flyrock.

5. Vibration Control:

• Vibration monitoring is essential to ensure that blast vibrations do not exceed established limits, which could cause damage to structures or infrastructure.

• Techniques like presplitting (creating a crack along the desired fracture plane before the main blast) can help control vibrations.

6. Environmental Impact:

- Blasting can generate dust, noise, and vibration, which may have environmental impacts. Dust suppression and noise control measures are often employed.

- Care is taken to prevent flyrock (pieces of rock thrown into the air during blasting) and to minimize the impact on nearby ecosystems and water bodies.

7. Post-Blast Inspection and Cleanup: After a blast, the area is inspected for safety, and any remaining rock fragments or debris are removed. Environmental mitigation measures, such as erosion control, may be employed.

Blasting is a highly specialized and regulated technique used in various industries to break hard materials safely and efficiently. It requires skilled professionals, adherence to strict safety protocols, and compliance with environmental regulations to ensure successful and responsible blasting operations.

16

TUNNELING

Tunneling is a construction technique that involves creating underground passageways, or tunnels, to provide transportation, utility access, or other forms of connectivity. Tunnels can vary in size, purpose, and complexity, and tunneling methods can be adapted to suit different geological conditions and project requirements. Here are the key aspects of tunneling:

1.   Purpose:

•   Transportation: Tunnels are commonly used to provide passage for roads, railways, subways, and pedestrian walkways, especially in urban areas where surface transportation is limited.

•   Utilities: Tunnels may house utility pipelines, cables, and conduits for purposes such as water supply, sewage, gas, electricity, and telecommunications.

•   Mining: In mining operations, tunnels are created to access mineral deposits and facilitate extraction.

•   Hydroelectric Projects: Tunnels can be integral to hydroelectric power generation by diverting water to turbines.

2. Tunneling Methods:

• **Conventional Tunneling:** Involves drilling and blasting (explosives) or mechanical excavation to create the tunnel. It is often used in rock or hard soil conditions.

• **Tunnel Boring Machines (TBMs):** TBMs are massive machines that mechanically excavate tunnels while simultaneously lining them with precast concrete or other materials. TBMs are suitable for a wide range of soil and rock conditions.

• **Cut-and-Cover:** This method involves excavating a trench, building the tunnel within it, and then covering it with backfill material. It is commonly used in urban areas.

• **Drill and Blast:** Involves drilling holes into rock or hard soil, filling them with explosives, and blasting to create the tunnel. This method is often used in mountainous terrain.

• **Shield Tunneling:** Shield machines are used to excavate tunnels in soft ground, such as clay or sand. They are equipped with a shield to support the tunnel face during excavation.

3. Geological and Environmental Considerations:

• The geological conditions of the area through which the tunnel passes are a critical factor in tunneling. The type of rock or soil, its stability, and groundwater levels affect the choice of tunneling method.

• Environmental considerations include minimizing disruption to ecosystems, managing water infiltration, and complying with regulations to protect natural resources.

4. Safety and Ventilation:

• Safety measures in tunneling include adequate ventilation to remove dust, gases, and exhaust fumes generated during excavation.

• Emergency egress routes and safety protocols are essential to protect workers and respond to emergencies.

5. Lining and Support:

• Tunnels often require lining or support systems to prevent collapses and maintain structural integrity. Linings can be made of concrete, steel, or other materials, depending on the tunnel's purpose and geological conditions.

6. Maintenance and Operation:

• Tunnels require regular maintenance to ensure safe operation. This includes monitoring structural integrity, maintaining lighting and ventilation systems, and addressing any damage or wear.

7.   Cost and Time:

•   Tunneling projects can be expensive and time-consuming due to the complexity of the work, geological challenges, and safety considerations.

8.   Surveying and Monitoring:

•   Precise surveying and monitoring are crucial to ensure that the tunnel alignment, grade, and dimensions meet design specifications and safety standards.

Tunneling is a versatile construction technique that plays a critical role in infrastructure development, transportation systems, and utility access. The choice of tunneling method depends on project requirements, geological conditions, and environmental considerations, and it often requires careful planning and engineering expertise.

17

DREDGING

Dredging is a process that involves the excavation and removal of sediments, debris, or underwater materials from bodies of water, such as rivers, lakes, harbors, and coastal areas. Dredging is essential for various purposes, including maintaining navigable waterways, deepening ports, managing sediment buildup, and environmental restoration. Here are the key aspects of dredging:

1.   Purpose:

•   Navigation: One of the primary purposes of dredging is to maintain or improve navigable waterways, ensuring that ships, boats, and vessels can safely travel through channels, ports, and harbors.

•   Port and Harbor Maintenance: Dredging is used to deepen ports, berths, and harbors to accommodate larger vessels, increasing a port's capacity and efficiency.

•   Sediment Removal: Dredging removes accumulated sediments, silt, sand, and debris from bodies of water, preventing shallowing and maintaining proper water depths.

• **Environmental Restoration:** In some cases, dredging is used for environmental purposes, such as restoring wetlands, removing contaminants, or creating habitat for wildlife.

• **Flood Control:** Dredging can help improve flood control by increasing the capacity of rivers and drainage channels to carry excess water during heavy rains or storms.

2. **Dredging Equipment:**

• **Dredgers:** Various types of dredgers are used for different applications. Cutter suction dredgers, trailing suction hopper dredgers, bucket dredgers, and clamshell dredgers are among the common types.

• **Pipelines:** Dredged material is often transported through pipelines to designated disposal areas or containment facilities.

• **Support Equipment:** Support vessels, tugboats, and equipment for sediment processing, dewatering, and disposal are used in conjunction with dredging operations.

3. **Dredging Methods:**

• **Mechanical Dredging:** This method involves using mechanical equipment, such as excavators or dredgers with buckets, to physically remove sediments and materials from the water bottom.

• **Hydraulic Dredging:** Hydraulic dredging uses water jets and suction to loosen and remove sediments, which are then transported through pipelines to a disposal location.

• **Environmental Dredging:** When dredging for environmental restoration, care is taken to avoid disturbing contaminated sediments or to remove them for proper treatment and disposal.

4.   Environmental Considerations:

• **Dredging projects must adhere to** environmental regulations and permits, especially when dealing with sensitive habitats, protected species, or potentially contaminated sediments.

• **Measures such as turbidity curtains, silt** fences, and sedimentation basins are often employed to minimize the environmental impact of dredging activities.

5.   Disposal of Dredged Material:

• **Dredged material can be disposed of in** various ways, including reclamation of land, creation of wetlands, disposal in designated areas, or treatment to remove contaminants.

• **In some cases, dredged material can be** beneficially used for beach nourishment, land reclamation, or construction fill.

6.   Monitoring and Surveying:

•   Accurate surveying and monitoring are critical to ensure that dredging activities achieve the required depths and remove the targeted volumes of material.

Dredging is a vital process for maintaining waterway infrastructure, ensuring safe navigation, and managing sediment buildup in bodies of water. However, it must be carried out with careful planning, environmental considerations, and adherence to regulations to minimize its impact on aquatic ecosystems and habitats.

18

HYDRO EXCAVATION

Hydro excavation, also known as vacuum excavation or hydrovac excavation, is a non-destructive digging method that uses pressurized water and a vacuum system to excavate soil and debris. It is commonly used in construction, utility installation, and various industries where precision excavation and the avoidance of underground utilities are critical. Here are the key aspects of hydro excavation:

1.   Purpose:

•   Safe Digging: Hydro excavation is used when traditional mechanical excavation methods, like digging with shovels or using heavy machinery, may pose a risk to underground utilities, pipelines, or sensitive infrastructure.

•   Utility Location: It is effective for exposing and verifying the location of buried utilities, such as gas lines, water pipes, telecommunications cables, and electrical conduits.

•   Precision Excavation: Hydro excavation allows for precise excavation with minimal disruption, making it suitable for jobs where accuracy is essential.

• **Trenching:** It is used for trenching, potholing, and slot trenching in various construction and maintenance projects.

2. **Equipment:**

• **Hydrovac Truck:** The primary equipment used in hydro excavation is a hydrovac truck, which consists of a high-pressure water system, a vacuum system, and a debris storage tank.

• **Water Jetting System:** The water system delivers pressurized water through a nozzle or wand, which breaks up the soil and debris, creating a slurry.

• **Vacuum System:** A powerful vacuum system simultaneously removes the slurry and debris from the excavation area and transports it to a storage tank.

• **Debris Storage Tank:** The debris is contained in a storage tank on the hydrovac truck until it can be properly disposed of or used as fill material.

3. **Operation:**

• The hydrovac operator directs a high-pressure water jet into the ground at the precise location where excavation is needed.

• The pressurized water loosens the soil and forms a slurry, which is immediately vacuumed up into the storage tank.

• The operator can control the excavation depth and precision, making it easy to avoid damaging underground utilities.

• The vacuumed material can be transported to a disposal site or used as backfill, depending on the project's requirements.

4. Safety and Environmental Benefits:

• Hydro excavation is a safer and more environmentally friendly method of excavation because it minimizes the risk of damaging underground utilities and reduces soil disturbance.

• It reduces the need for disruptive and potentially hazardous mechanical digging methods, such as backhoes or excavators.

• The process is less likely to release harmful chemicals or contaminants into the environment.

5. Accuracy and Efficiency:

• Hydro excavation is highly accurate, allowing for precision digging and minimal over-excavation.

• It is efficient and can significantly reduce the time and labor required for excavation tasks.

6.   **Applications:**

•   **Hydro excavation is commonly used in utility location and verification, slot trenching, daylighting (exposing buried utilities), potholing (digging test holes), and general excavation where precision is required.**

7.   **Regulations and Permits:**

•   **Depending on the location and type of work, hydro excavation may require permits and compliance with local regulations and safety standards.**

Hydro excavation is a valuable method for safe, precise, and non-destructive digging. It plays a critical role in preventing damage to underground utilities, improving job site safety, and minimizing the environmental impact of excavation projects.

19

PNEUMATIC EXCAVATION

Pneumatic excavation, also known as air excavation, is a non-destructive digging method that uses compressed air to break up and remove soil, debris, and other materials from an excavation site. This technique is particularly useful in situations where mechanical excavation methods or traditional digging could damage underground utilities or structures. Here are the key aspects of pneumatic excavation:

1.   Purpose:

•   Safe Digging: Pneumatic excavation is used when conventional excavation methods may pose a risk to underground utilities, pipelines, or sensitive infrastructure.

•   Utility Location: It is effective for exposing and verifying the location of buried utilities, such as gas lines, water pipes, telecommunications cables, and electrical conduits.

•   Precision Excavation: Pneumatic excavation allows for precise digging with minimal disruption, making it suitable for jobs where accuracy is essential.

• **Trenching:** It can be used for trenching, potholing, and slot trenching in various construction and maintenance projects.

2. **Equipment:**

• **Pneumatic Excavator:** The primary equipment used in pneumatic excavation is a pneumatic excavator, which consists of a compressed air system, a vacuum system, and a debris collection tank.

• **Compressed Air System:** A high-pressure air compressor generates compressed air that is delivered through a nozzle or lance.

• **Nozzle or Lance:** The compressed air is directed into the ground at the precise location where excavation is needed.

• **Vacuum System:** A vacuum system simultaneously removes the loosened soil and debris from the excavation area and transports it to a collection tank.

• **Debris Collection Tank:** The debris is stored in a collection tank on the pneumatic excavator until it can be properly disposed of or used as backfill material.

3. **Operation:**

• The pneumatic excavator operator directs the high-pressure air into the ground, causing the soil and debris to become dislodged.

• The vacuum system immediately removes the dislodged material and transports it into the collection tank.

• The operator can control the excavation depth and precision, making it easy to avoid damaging underground utilities or structures.

• The vacuumed material can be transported to a disposal site or used as backfill, depending on the project's requirements.

4. Safety and Environmental Benefits:

• Pneumatic excavation is a safer and more environmentally friendly method of digging because it minimizes the risk of damaging underground utilities and reduces soil disturbance.

• It reduces the need for disruptive and potentially hazardous mechanical digging methods, such as backhoes or excavators.

• The process is less likely to release harmful chemicals or contaminants into the environment.

5. Accuracy and Efficiency:

- Pneumatic excavation is highly accurate, allowing for precision digging and minimal over-excavation.

- It is efficient and can significantly reduce the time and labor required for excavation tasks.

6. Applications:

- Pneumatic excavation is commonly used in utility location and verification, daylighting (exposing buried utilities), potholing (digging test holes), slot trenching, and general excavation where precision is required.

7. Regulations and Permits:

- Depending on the location and type of work, pneumatic excavation may require permits and compliance with local regulations and safety standards.

Pneumatic excavation is a valuable method for safe, precise, and non-destructive digging. It plays a critical role in preventing damage to underground utilities, improving job site safety, and minimizing the environmental impact of excavation projects.

20

ROCK EXCAVATION

Rock excavation is the process of breaking, loosening, or removing solid rock from the Earth's surface or subsurface for various construction, mining, and geological purposes. It involves the use of specialized equipment and techniques to deal with the hardness and strength of rock formations. Here are the key aspects of rock excavation:

1.   Purpose:

•   Construction: Rock excavation is often necessary for construction projects, such as foundations, tunnels, roads, bridges, and building sites, where the presence of solid rock must be removed or altered.

•   Mining: In the mining industry, rock excavation is essential for accessing valuable minerals or ore deposits buried within solid rock formations.

•   Tunneling: Construction of tunnels for transportation, utilities, or mining operations often involves extensive rock excavation.

•   Landscaping: In some landscaping projects, large rocks or boulders need to be removed or reshaped to create desired aesthetics.

2.   **Types of Rock Excavation:**

•   **Mechanical Excavation:** Mechanical methods involve the use of heavy machinery such as excavators, backhoes, and bulldozers equipped with specialized attachments like rock breakers, hydraulic hammers, or ripper teeth to break and remove rock.

•   **Blasting:** Explosives are used to break and fragment rock in controlled blasting operations. This method is common in mining, tunneling, and large-scale rock removal projects.

•   **Chemical and Thermal Methods:** Chemical agents or heat may be used to weaken or fracture the rock, making it easier to remove.

3.   **Rock Hardness and Composition:**

•   The hardness, strength, and composition of the rock play a crucial role in selecting the appropriate excavation method.

•   Harder rocks like granite and basalt may require more specialized equipment and techniques than softer rocks like limestone or sandstone.

4.   **Blasting in Rock Excavation:**

•   Controlled blasting involves drilling holes into the rock, inserting explosives, and triggering a controlled detonation to fracture the rock.

• Blasting patterns and drilling techniques are carefully designed to achieve the desired fragmentation without causing excessive damage or vibration to nearby structures or geological formations.

• Safety measures and monitoring are essential during blasting operations to protect workers and the environment.

5. Safety and Environmental Considerations:

• Rock excavation can be hazardous due to the risk of falling rocks, cave-ins, and exposure to dust and hazardous materials.

• Safety protocols, protective gear, and proper equipment are critical to minimize risks.

• Environmental impact assessments and mitigation measures are often required to prevent adverse effects on nearby ecosystems and water bodies.

6. Monitoring and Surveying:

• Accurate surveying and monitoring are essential to ensure that the excavation is progressing according to design specifications.

• Instruments such as inclinometers, seismographs, and surveying equipment are used to measure rock stability and ground movements.

7.	**Waste Management and Disposal:**

•	Excavated rock material may be used as fill material in construction or landscaping projects, or it may need to be transported and disposed of in designated areas, such as landfills or quarries.

Rock excavation requires careful planning, engineering expertise, and adherence to safety and environmental regulations. The choice of excavation method and equipment is influenced by the specific geological conditions, project goals, and the type of rock being excavated. Proper planning and execution are essential to ensure the success of rock excavation projects.

21

EXCAVATORS

Excavators, also known as diggers or hydraulic excavators, are versatile heavy construction machines designed for digging, earthmoving, and material handling tasks. They are commonly used in various construction, mining, demolition, and excavation projects. Excavators are known for their mobility, power, and ability to handle a wide range of attachments. Here are the key aspects of excavators:

1.   Components and Structure:

•   Cab: Excavators have a cab where the operator sits. The cab is equipped with controls, instruments, and visibility features to ensure safe and efficient operation.

•   Undercarriage: The undercarriage consists of tracks or wheels that allow the excavator to move across different terrains. Tracks provide stability and distribute the machine's weight over a larger area, making them suitable for rough or soft ground.

•   Boom: The boom is the main arm of the excavator and connects to the cab. It provides reach and elevation for digging or lifting operations.

• **Stick (Arm):** The stick, also known as the arm, attaches to the end of the boom and provides additional length and control for digging and excavation.

• **Bucket:** The bucket is attached to the end of the stick and is used for digging, scooping, and loading materials.

• **Hydraulic System:** Excavators are powered by a hydraulic system that controls the movement of the boom, stick, and bucket. It provides the necessary force for digging and lifting operations.

2. **Operation:**

• Excavators are operated by a trained operator who uses hand and foot controls inside the cab.

• The operator can control the movement of the machine, including rotation, elevation, and extension of the boom, stick, and bucket.

• Excavators are highly maneuverable, capable of rotating 360 degrees on their undercarriage, which allows them to access tight or confined spaces.

3. **Attachments:**

• Excavators can be equipped with various attachments to perform different tasks, such as:

- **Grapple:** Used for handling and sorting materials.

- **Auger:** Used for drilling holes in the ground.

- **Hammer:** Used for breaking up concrete or rock.

- **Ripper:** Used for breaking up hard soil or rock.

- **Thumb:** Provides a secure grip for handling irregularly shaped objects.

- **Compactor:** Used for compacting soil or other materials.

4.   **Applications:**

- **Excavators** are used in a wide range of applications, including:

- **Earthmoving:** Excavating and moving soil, rocks, and debris on construction sites.

- **Digging:** Digging trenches, foundations, and utility lines.

- **Demolition:** Demolishing buildings and structures.

- **Mining:** Extracting minerals and ore from mines.

• **Forestry: Clearing land and handling timber.**

• **Landscaping: Shaping and grading terrain.**

• **Material Handling: Loading and unloading materials from trucks or conveyor belts.**

5. **Maintenance:**

• **Regular maintenance is essential to keep excavators in good working condition.**

• **Maintenance tasks include checking fluid levels, inspecting tracks or wheels, lubricating components, and inspecting hydraulic systems.**

• **Operators should also perform daily pre-operation inspections to ensure safety and reliability.**

6. **Safety:**

• **Excavator operators must receive proper training to operate the equipment safely.**

• **Safety features, such as rollover protection structures (ROPS) and falling object protection structures (FOPS), are often incorporated into excavator designs.**

• **Safety protocols, including the use of seat belts and adherence to site-specific safety regulations, are crucial.**

A SHORT BRIEF OF EXCAVATION
BY RAMCHANDRA SHRIVASTAV

Excavators are vital machines in the construction and excavation industry due to their versatility, power, and efficiency. They are available in various sizes and configurations to suit different project requirements, from small-scale landscaping to large-scale construction and mining operations.

22

BACKHOES

A backhoe, also known as a backhoe loader, is a versatile heavy equipment machine commonly used in construction, excavation, and landscaping projects. It combines the capabilities of both a tractor (with wheels or tracks) and a front loader with a bucket and a rear excavator with a digging bucket or attachment. Backhoes are known for their ability to perform a wide range of tasks, making them a popular choice in various industries. Here are the key aspects of backhoes:

1. Components:

• Front Loader: The front part of the backhoe features a loader bucket attached to the front arms or boom. The loader bucket is used for scooping, lifting, and carrying materials, such as dirt, gravel, sand, or debris.

• Rear Excavator: The rear section of the backhoe has a hydraulic arm with a digging bucket or other attachments. This part is used for digging trenches, excavating, and performing precision excavation tasks.

• Cab: Backhoes are typically equipped with an enclosed cab where the operator sits. The cab contains controls, instruments, and safety features to ensure efficient and safe operation.

• Stabilizers: Extendable stabilizers or outriggers are often deployed from the backhoe's rear to provide stability when performing digging operations.

• Hydraulic System: Backhoes are powered by a hydraulic system that controls the movement of the loader and excavator arms, the bucket attachments, and other functions.

2. Operation:

• A trained operator uses the controls in the cab to manipulate the loader and excavator arms, as well as the bucket attachments.

• The operator can switch between loader and excavator modes as needed, allowing for a seamless transition between digging, lifting, and loading tasks.

• Backhoes are highly maneuverable, with the ability to rotate 180 degrees on their axis, making them suitable for confined spaces.

3. Applications:

• Backhoes are used in a wide range of applications, including:

A SHORT BRIEF OF EXCAVATION
BY RAMCHANDRA SHRIVASTAV

• **Excavation: Digging trenches, foundations, and holes for various purposes, such as utility installation and landscaping.**

• **Loading: Scooping and transporting materials, such as soil, gravel, or construction debris.**

• **Material Handling: Lifting and moving heavy objects or pallets.**

• **Demolition: Breaking up concrete, asphalt, or small structures.**

• **Landscaping: Grading, leveling, and shaping terrain for landscaping projects.**

• **Snow Removal: Clearing snow from roads and parking lots using a snowplow attachment.**

4. **Attachments:**

• **Backhoes can be equipped with various attachments, expanding their versatility. Some common attachments include augers, hammers, grapples, and thumbs.**

5. **Maintenance:**

• **Routine maintenance is crucial to ensure the backhoe operates safely and efficiently. Maintenance tasks include checking fluid levels, inspecting and lubricating components, and ensuring safety features are functional.**

6.   **Safety:**

•   **Backhoe operators should receive proper training to operate the equipment safely.**

•   **Safety features such as rollover protection structures (ROPS) and falling object protection structures (FOPS) are often incorporated into backhoe designs.**

•   **Operators should use seat belts and adhere to site-specific safety regulations.**

Backhoes are versatile and valuable machines in the construction and excavation industry, as well as in agricultural and utility maintenance tasks. They are known for their flexibility, allowing operators to perform a wide range of tasks with a single piece of equipment.

23

BULLDOZERS

Bulldozers, often referred to as dozers, are heavy construction machines characterized by a large, powerful blade at the front of the vehicle and a tracked or wheeled undercarriage. They are commonly used in various construction, mining, and earthmoving applications due to their exceptional pushing and leveling capabilities. Here are the key aspects of bulldozers:

1.   Components and Structure:

•   Blade: The most prominent feature of a bulldozer is its front-mounted blade, which can vary in size and type. Blades can be straight, angled, or U-shaped, depending on the specific task.

•   Undercarriage: Bulldozers can have either tracks or wheels. Tracks provide better traction and are suitable for soft or uneven terrain, while wheeled bulldozers are more maneuverable and suitable for harder surfaces.

•   Engine: Bulldozers are equipped with powerful diesel engines that provide the necessary horsepower to perform heavy-duty tasks.

• Cab: The operator sits inside an enclosed cab equipped with controls, instruments, and safety features to ensure efficient and safe operation.

• Ripper: Some bulldozers have a rear-mounted ripper attachment used for breaking up hard or compacted soil or rock.

2. Operation:

• Bulldozer operators use the controls in the cab to manipulate the blade and the movement of the machine.

• The blade can be raised, lowered, tilted, and angled to perform a variety of tasks, including pushing, leveling, grading, and excavating.

• Tracks or wheels provide mobility and stability as the bulldozer moves across the work site.

3. Applications:

• Bulldozers are used in a wide range of applications, including:

• Earthmoving: Moving and leveling large volumes of soil, rock, or debris on construction sites.

• Site Preparation: Clearing land, removing vegetation, and preparing the ground for construction projects.

- **Road Construction:** Building and maintaining roads, highways, and embankments.

- **Mining:** Excavating and transporting materials in mining operations.

- **Land Reclamation:** Filling and shaping land for development or environmental restoration.

- **Forestry:** Clearing and grading land for forestry operations.

- **Snow Removal:** Clearing snow from roads and large areas during winter.

4. Attachments:

- Bulldozers can be equipped with various attachments, such as rippers, winches, and brush guards, to enhance their versatility and performance in specific tasks.

5. Maintenance:

- Routine maintenance is essential to ensure the bulldozer operates safely and efficiently. Maintenance tasks include checking fluid levels, inspecting and lubricating components, and ensuring safety features are functional.

6. Safety:

- Bulldozer operators should receive proper training to operate the equipment safely.

•   Safety features, including rollover protection structures (ROPS) and falling object protection structures (FOPS), are often incorporated into bulldozer designs.

•   Operators should use seat belts and adhere to site-specific safety regulations.

Bulldozers are known for their power, durability, and versatility in tackling a wide range of construction and earthmoving tasks. They are essential machines in various industries due to their ability to move and shape large quantities of material efficiently. The choice between tracked and wheeled bulldozers depends on the specific requirements of the job and the terrain conditions.

24

DRAGLINE EXCAVATORS

Dragline excavators, often referred to simply as draglines, are massive and specialized pieces of heavy equipment used primarily in large-scale surface mining and civil engineering projects. They are known for their ability to efficiently excavate and move massive quantities of earth, overburden, or materials. Draglines are characterized by their long boom and bucket, which are typically suspended from a large counterweighted structure. Here are the key aspects of dragline excavators:

1.   Components and Structure:

•   Boom: The boom is a long, hinged, and articulated arm that extends outward from the dragline's central structure. It supports the bucket and allows for digging and swinging operations.

•   Bucket: The bucket is a large container attached to the end of the boom. It is used for excavating, scooping, and lifting materials.

•   Counterweight: To maintain balance and stability, draglines are equipped with a massive counterweight structure typically located at the rear of the machine.

- **Undercarriage:** Some draglines are mounted on tracks or crawlers, while others may have a static base with wheels or a setup on floating pontoons, depending on the specific application.

- **Operator's Cab:** Draglines have an operator's cab that provides visibility and control over the machine's movements and operations.

2. **Operation:**

- Draglines operate by using a hoisting system to raise and lower the bucket and a system of cables and ropes to move the bucket horizontally.

- The operator controls the dragline from the cab, using various levers, pedals, and controls to perform digging, swinging, and dumping operations.

- Draglines are capable of digging at impressive depths and distances due to the length of their boom and the counterweight system.

3. **Applications:**

- **Mining:** Draglines are commonly used in large-scale surface mining operations, such as coal, iron ore, and phosphate mining, for removing overburden (soil, rock, and other materials) to expose valuable mineral deposits.

- **Civil Engineering:** They are also used in civil engineering projects, including dam construction, canal excavation, and port development, where massive earthmoving tasks are required.

• Environmental Remediation: In some cases, draglines are employed for environmental projects like land reclamation and wetland restoration.

4. Advantages:

• High Production Rates: Draglines are known for their high production rates and efficiency in moving large quantities of material in a relatively short amount of time.

• Selective Mining: They can selectively remove overburden, preserving valuable mineral deposits.

• Reduced Ground Pressure: Some draglines are mounted on tracks or pontoons, which help distribute the machine's weight over a larger area and minimize ground pressure in sensitive environments.

5. Size and Capacity:

• Draglines come in various sizes, with bucket capacities ranging from several cubic meters to over 100 cubic meters.

• The largest draglines in the world are capable of excavating tens of thousands of cubic meters of material in a single day.

6.   Maintenance and Safety:

•   Draglines require regular maintenance to ensure safe and efficient operation.

•   Safety is a paramount concern due to the size and complexity of these machines, and operators receive extensive training to operate them safely.

Dragline excavators are among the most powerful and productive earthmoving machines in the world. They play a crucial role in the mining and construction industries by enabling the removal of massive volumes of material with precision and efficiency.

25

TRENCHERS

Trenchers are specialized construction machines designed for excavating trenches, which are narrow, long, and typically shallow channels dug into the ground. Trenches serve various purposes, including utility installation (such as for water, gas, or electrical lines), drainage, irrigation, and cable laying. Trenchers come in various sizes and configurations to suit different trenching needs. Here are the key aspects of trenchers:

1.   Types of Trenchers:

•   Chain Trenchers: These trenchers use a digging chain or belt equipped with digging teeth or blades to cut through the soil. Chain trenchers are commonly used for a wide range of trenching tasks and are available in different sizes and configurations.

•   Wheel Trenchers: Wheel trenchers feature a rotating wheel with digging teeth or buckets that cut through the ground. They are often used for shallow trenching and are suitable for compacted or rocky soil conditions.

•   Micro Trenchers: Micro trenchers are smaller, specialized machines designed for creating narrow and shallow trenches, often used for installing fiber optic cables and small utility lines in urban areas.

• **Rock Wheel Trenchers:** These trenchers are designed for trenching through rocky terrain and feature robust cutting wheels with carbide teeth to handle hard materials.

• **Chain-Boom Trenchers:** These trenchers combine a chain system with a boom that allows for deeper and wider trenching than traditional chain trenchers.

• **Hydro Trenchers:** Hydro trenchers use pressurized water to cut through soil and create trenches. They are often used in areas with sensitive utilities to avoid damage.

2. **Components:**

• **Digging Mechanism:** The digging mechanism varies depending on the type of trencher and may involve a chain, wheel, or other digging component.

• **Conveyor System:** Trenchers typically include a conveyor system to remove excavated soil and debris from the trench and deposit it to the side.

• **Operator's Station:** Trenchers have an operator's station where the operator controls the machine and monitors the trenching process.

• **Engine: Trenchers are powered by engines that provide the necessary power for digging and moving the machine.**

3. Operation:

• **Trenchers are operated by a trained operator who controls the machine from the operator's station.**

• **The operator can adjust the depth and width of the trench, as well as the rate of excavation, depending on the specific requirements of the project.**

• **Trenchers are typically equipped with safety features and controls to ensure safe operation.**

4. Applications:

• **Utility Installation: Trenchers are commonly used to dig trenches for water, sewer, gas, and electrical utility lines.**

• **Drainage and Irrigation: Trenchers can create trenches for drainage systems, agricultural irrigation, and landscaping projects.**

• **Cable Laying: Trenchers are used in the telecommunications industry to lay cables for internet, phone, and television services.**

• **Pipeline Installation: They are also employed for pipeline installation and repair in the oil and gas industry.**

5. **Maintenance and Safety:**

• **Regular maintenance is essential to keep trenchers in good working condition.**

• **Safety measures, including operator training and adherence to safety regulations, are crucial to prevent accidents during trenching operations.**

Trenchers are valuable tools in the construction and utility industries, allowing for efficient and precise trenching, which is critical for infrastructure development and maintenance. The choice of trencher type depends on the specific trenching requirements and site conditions.

26

HYDRAULIC SHOVELS

Hydraulic shovels, also known as hydraulic excavators or simply excavators, are heavy construction machines designed for digging, excavating, and lifting operations. They are a common sight on construction sites, mining operations, and in various earthmoving applications. Hydraulic shovels are characterized by their versatile and precise digging capabilities, which are achieved through a hydraulic system that powers their movements. Here are the key aspects of hydraulic shovels:

1.   Components and Structure:

•   Boom: The boom is a large, hinged arm that extends from the excavator's body. It provides vertical reach and lifting capacity.

•   Stick (Arm): The stick, also known as the arm, attaches to the end of the boom and provides horizontal reach.

•   Bucket: The bucket is attached to the end of the stick and is used for digging, scooping, and loading materials.

•   Undercarriage: Excavators may have either tracks or wheels for mobility. Track-mounted excavators provide better stability and weight distribution for softer terrain, while wheeled excavators are more maneuverable on harder surfaces.

• **Engine:** Hydraulic shovels are powered by powerful engines that drive the hydraulic system and provide the necessary horsepower for digging and lifting.

• **Operator's Cab:** The operator's cab is equipped with controls, instrumentation, and visibility features for efficient and safe operation.

2. **Hydraulic System:**

• Hydraulic shovels rely on a hydraulic system to control their movements. Hydraulic fluid is pressurized and distributed to hydraulic cylinders to power the excavator's various functions.

• The hydraulic system allows precise control of the boom, stick, and bucket for digging, swinging, and lifting operations.

3. **Operation:**

• Trained operators use the controls in the cab to manipulate the boom, stick, and bucket, as well as to control the movement of the machine.

• Hydraulic shovels are highly versatile and can perform a wide range of tasks, including digging trenches, excavating foundations, loading trucks, and precision digging in confined spaces.

4.   Applications:

•   Hydraulic shovels are used in a variety of applications, including:

•   Construction: Digging foundations, trenches, and utility lines; removing debris; and performing grading and landscaping tasks.

•   Mining: Excavating overburden, ore, and other materials in open-pit mining operations.

•   Demolition: Breaking up and removing concrete, structures, and debris.

•   Forestry: Clearing land, handling timber, and assisting in forestry operations.

•   Material Handling: Loading and unloading materials from trucks, stockpiles, or conveyor systems.

5.   Attachments:

•   Hydraulic shovels can be equipped with various attachments, such as grapples, hydraulic hammers, rippers, and thumbs, to enhance their versatility and adapt to different tasks.

6.   Maintenance:

•   **Regular maintenance is crucial to ensure the hydraulic shovel operates safely and efficiently. Maintenance tasks include checking fluid levels, inspecting and lubricating components, and ensuring safety features are functional.**

7.   Safety:

•   **Operators of hydraulic shovels should receive proper training to operate the equipment safely.**

•   **Safety features, including rollover protection structures (ROPS) and falling object protection structures (FOPS), are often incorporated into excavator designs.**

•   **Operators should use seat belts and adhere to site-specific safety regulations.**

Hydraulic shovels are powerful and versatile machines capable of handling a wide range of excavation and earthmoving tasks. They are essential in construction, mining, and other industries where precise and efficient digging and lifting operations are required. The choice between tracked and wheeled excavators depends on the specific job requirements and site conditions.

27

MANUAL SHOVELING

Manual shoveling refers to the process of digging, lifting, and moving materials or soil using hand tools such as shovels and spades. This labor-intensive method is commonly used in various industries, including construction, agriculture, gardening, landscaping, and snow removal. Here are the key aspects of manual shoveling:

1.   Tools:

•   Shovel: Shovels come in various types and sizes, each designed for specific tasks. Common types include square-nose shovels for digging and lifting, round-nose shovels for digging in softer soil, and snow shovels with wide blades for clearing snow.

•   Spade: Spades have a flat, pointed blade and are typically used for digging trenches, edging, and more precise excavation.

•   Hand Trowel: Smaller hand trowels are used for tasks like planting, weeding, and transplanting in gardening and landscaping.

•   Pitchfork: Pitchforks have multiple prongs and are used for lifting and turning materials like hay, straw, or compost.

2.   Operation:

•   Manual shoveling involves physically digging into the ground or material, lifting the load onto the shovel or spade, and then transferring it to another location.

•   The operator uses their strength and body mechanics to perform the digging and lifting tasks.

•   The load is carried or pushed to the desired location, where it can be dumped or spread.

3.   Applications:

•   Construction: Manual shoveling is often used in construction for tasks such as excavating small areas, spreading gravel, or moving concrete.

•   Gardening and Landscaping: Gardeners and landscapers use shovels and spades for planting, digging holes, edging, and moving soil or mulch.

•   Agriculture: Farmers use shovels and pitchforks for tasks like turning compost, spreading manure, or digging trenches for irrigation.

•   Snow Removal: Snow shovels are specifically designed for clearing snow from driveways, sidewalks, and walkways.

4.   Safety Considerations:

• Manual shoveling can be physically demanding and may lead to injuries or strains if not performed correctly.

• Proper body mechanics are essential to reduce the risk of injuries. This includes using the legs to lift and keeping the back straight.

• Adequate rest breaks and hydration are important during strenuous manual shoveling tasks.

• Wearing appropriate personal protective equipment, such as gloves, steel-toed boots, and eye protection, can enhance safety.

5. Efficiency:

• Manual shoveling is often slower and less efficient than using powered equipment, especially for large-scale excavation or material handling tasks.

• Efficiency can be improved with proper technique and using the right tool for the job.

6. Environmental Impact:

• Manual shoveling is generally environmentally friendly as it does not involve the emissions associated with powered equipment.

• It is a sustainable method of moving materials in many small-scale applications.

A SHORT BRIEF OF EXCAVATION
BY RAMCHANDRA SHRIVASTAV

Manual shoveling is a fundamental and versatile method for digging, lifting, and moving materials, and it remains an important tool in various industries. While it can be physically demanding, proper technique and safety precautions can help reduce the risk of injury and ensure efficient and effective manual shoveling tasks.

28

PICKAXES AND MATTOCKS

Pickaxes and mattocks are hand tools designed for digging, breaking up soil, rock, and other materials. They are commonly used in construction, agriculture, gardening, mining, and landscaping. While they share some similarities, they have distinct features and purposes. Here are the key aspects of pickaxes and mattocks:

Pickaxe:

1.   Design:

•   A pickaxe consists of a long handle made of wood or fiberglass, with a metal head at one end.

•   The head typically has two ends: a pointed pick and an adze or chisel-like blade.

2.   Purpose:

•   The pointed pick is designed for breaking up hard materials such as rock, concrete, or compacted soil. It can be driven into the material and then leveraged to break it apart.

- The adze or chisel end is used for digging and prying. It is effective for removing loose soil, creating holes, or shaping surfaces.

3.   Applications:

- Pickaxes are commonly used in construction and mining for breaking rock and concrete.

- In gardening and landscaping, they can be used for digging holes, loosening soil, and removing roots or stubborn rocks.

Mattock:

1.   Design:

- A mattock has a handle, similar to that of a pickaxe, but with a different head configuration.

- The head features an adze or chisel-like blade on one side and a pick or hoe-like blade on the other side.

- The pick blade is typically flat and broad, while the adze blade is narrower and curved.

2.   Purpose:

- The adze blade is designed for digging, cutting through roots, and shaping soil or surfaces. It can be effective for trenching and cultivating.

• The pick blade, often referred to as the grub hoe or grubbing blade, is used for breaking up soil, especially in clay-heavy or compacted soils. It can also be used for weeding and digging.

3. Applications:

• Mattocks are frequently used in agriculture and gardening for preparing soil, digging trenches, and clearing land.

• They are useful for tasks such as removing tree stumps, digging irrigation channels, and working in dense or rocky soil.

Common Features:

• Both pickaxes and mattocks have handles that allow the user to swing the tool to apply force effectively.

• They are manually operated, requiring physical effort from the user.

• Proper safety measures, including the use of eye protection and gloves, should be followed when using these tools to prevent accidents or injuries.

Choosing Between Pickaxes and Mattocks:

• The choice between a pickaxe and a mattock depends on the specific task and the type of material you are working with. If you primarily need to break up hard materials like rock or concrete, a pickaxe is more suitable. If you need to dig, cultivate, or work with softer soils, a mattock may be a better choice due to its versatile head configuration.

Both pickaxes and mattocks are valuable hand tools for various outdoor tasks that involve digging, breaking, and shaping soil or materials. The choice between them should be based on the requirements of the job at hand.

29

MANUAL TRENCHING

Manual trenching refers to the process of digging trenches by hand using basic hand tools. This labor-intensive method is typically employed for smaller-scale trenching projects, such as those in gardening, landscaping, plumbing, or utility installation. While manual trenching is slower than using powered equipment like trenchers or excavators, it can be cost-effective for smaller tasks and in situations where access is limited. Here are the key aspects of manual trenching:

Tools:

1.   Shovel: A shovel with a square-nose or round-nose blade is commonly used for digging trenches. Square-nose shovels are effective for straight-sided trenches, while round-nose shovels are useful for curved or irregular trenches.

2.   Spade: A spade, which has a flat, pointed blade, can be used for more precise trenching, especially when the trench needs to have straight, clean edges.

3.   Pickaxe or Mattock: A pickaxe or mattock may be used to break up hard or compacted soil, rocks, or roots encountered during trenching. These tools are particularly handy for deeper trenches.

4.   Trowel: A small hand trowel is useful for fine-tuning the trench's bottom and sides, especially in gardening or landscaping projects.

Operation:

1.   Planning: Before starting, carefully plan the trench's location, depth, width, and path to ensure it meets your project's requirements.

2.   Marking: Mark the trench's path on the ground using stakes, string, or spray paint to serve as a guide while digging.

3.   Digging: Use the appropriate hand tool (shovel, spade, pickaxe, or mattock) to start digging along the marked path. Begin at one end and work your way to the other.

4.   Depth and Width: Regularly check the trench's depth and width using a measuring tape or other suitable measuring tools to ensure it meets your project's specifications.

5.   Sloping: If your trench needs sloping (to ensure proper drainage, for example), dig the trench with a slight slope from the starting point to the endpoint.

6.   Removing Excavated Soil: As you dig, place the excavated soil in a wheelbarrow, bucket, or tarp for removal from the trench area.

7.   **Safety:** Take care to avoid overexertion and back strain while digging. Use proper body mechanics by bending your knees and lifting with your legs. Take regular breaks to rest and stay hydrated.

Applications:

•   Manual trenching is suitable for various tasks, such as:

•   Installing utility lines (water, gas, electrical, or irrigation).

•   Creating garden beds or planting rows.

•   Installing edging or drainage systems.

•   Digging trenches for foundation footings or landscaping features.

Efficiency:

•   Manual trenching is slower and more physically demanding compared to using powered equipment like trenchers or excavators.

•   The efficiency of manual trenching depends on factors such as soil type, depth, width, and the operator's strength and skill.

Safety Considerations:

• Manual trenching can be physically demanding. Use proper lifting techniques to avoid strain or injury.

• Wear appropriate personal protective equipment, including gloves, sturdy boots, and eye protection.

• Be cautious when digging near existing utilities to avoid damage or injury.

• Stay hydrated and take breaks to prevent overheating or exhaustion.

Manual trenching can be a cost-effective and practical solution for small-scale trenching projects. However, for larger or more extensive trenching tasks, or when working with challenging soil conditions, powered equipment may be a more efficient and practical choice.

30

PARTICLE SIZE ANALYSIS

Particle size analysis is a laboratory technique used to determine the size distribution of particles in a sample of material. This analysis is crucial in various fields, including geology, environmental science, pharmaceuticals, agriculture, and materials science. It provides valuable information about the physical properties of particles, which can influence material behavior, processing, and performance. Here are the key aspects of particle size analysis:

1. Importance:

• Quality Control: Particle size analysis is essential for quality control in manufacturing processes, ensuring that products meet specifications and performance standards.

• Research and Development: It aids in the development of new materials and formulations by understanding how particle size affects properties such as strength, reactivity, and solubility.

• Environmental Monitoring: Particle size analysis can help assess the environmental impact of particulate matter, such as pollutants, dust, or sediment in water bodies.

2. Methods:

•　　Sieve Analysis: In this method, particles are separated into size fractions using a series of stacked sieves with progressively smaller openings. It's suitable for coarse particles like sands and gravels.

•　　Laser Diffraction: Laser diffraction instruments use laser light scattering to determine particle size distribution. This method covers a wide size range, from submicron to millimeters, making it versatile for many materials.

•　　Sedimentation: Sedimentation methods, such as pipette or hydrometer analysis, rely on the settling rate of particles in a liquid to estimate particle size distribution.

•　　Microscopy: Microscopic analysis involves directly observing and measuring particles under a microscope. It's particularly useful for very small particles, but it's time-consuming.

•　　Coulter Counter: This method relies on electrical impedance measurements as particles pass through a small aperture. It's suitable for submicron to micron-sized particles.

3. Sample Preparation:

•　　Proper sample preparation is essential to ensure accurate results. It includes processes like drying, homogenizing, and dispersing the sample to ensure that particles are evenly distributed and representative of the material.

4. Data Presentation:

• Particle size distribution data is typically presented in the form of a histogram or a cumulative distribution curve, showing the percentage of particles within specific size ranges.

5. Applications:

• Mining and Geology: Particle size analysis helps in mineral processing and soil classification.

• Pharmaceuticals: It's critical in drug formulation to ensure consistent dosage and efficacy.

• Agriculture: Understanding soil particle size distribution is vital for crop management and irrigation.

• Environmental Science: Particle size analysis is used in air quality monitoring, sediment transport studies, and wastewater treatment.

• Material Science: It informs the development of materials like ceramics, polymers, and composites.

6. Factors Affecting Particle Size:

• Material Composition: Different materials have different size distributions based on their inherent properties.

•   **Processing Methods:** Mechanical processing, such as grinding or milling, can alter particle size.

•   **Environmental Factors:** Natural processes like weathering can affect particle size distribution in soils.

7. Quality Control:

•   Particle size analysis is an integral part of quality control in industries such as construction, pharmaceuticals, and food processing, ensuring that products meet specifications.

8. Regulatory Compliance:

•   Some industries, like pharmaceuticals and food, have strict regulatory requirements for particle size distribution to ensure product safety and effectiveness.

Particle size analysis is a versatile and valuable technique that provides insights into the behavior and properties of materials. It plays a critical role in various industries and scientific disciplines, contributing to product development, quality control, and research efforts.

31

MOISTURE CONTENT ANALYSIS

Moisture content analysis, also known as moisture determination or moisture testing, is a laboratory technique used to measure the amount of moisture or water content in a sample of material. This analysis is essential in various fields, including food processing, agriculture, construction, environmental science, and materials science. Accurate moisture content analysis is crucial because the moisture content of a material can significantly impact its quality, stability, and performance. Here are the key aspects of moisture content analysis:

1. Importance:

• Quality Control: Moisture content analysis is vital for ensuring product quality and consistency in various industries, such as food production and pharmaceuticals.

• Research and Development: It aids in research efforts to optimize processes and develop new products by understanding how moisture affects material properties and behavior.

- **Environmental Monitoring:** It's used to assess soil moisture, which is critical for agriculture, forestry, and land management.

- **Construction:** In construction materials like concrete and gypsum, moisture content can affect strength and durability.

2. Methods:

- **Gravimetric Method:** This is the most common method, involving the precise weighing of a sample before and after drying it to evaporate the moisture. The difference in weight represents the moisture content.

- **Karl Fischer Titration:** This method is specifically designed for measuring trace amounts of moisture in various substances. It involves a chemical reaction with the moisture, and the amount of reagent consumed is used to calculate moisture content.

- **Near-Infrared (NIR) Spectroscopy:** NIR spectroscopy can be used for non-destructive moisture content analysis. It relies on the absorption of near-infrared light by water molecules.

- **Drying Oven Method:** This method involves placing a sample in a drying oven at a specified temperature and time until it reaches a constant weight.

3. Sample Preparation:

• Proper sample preparation is crucial to ensure accurate results. It includes homogenizing the sample, ensuring it is representative of the material, and accurately weighing it.

4. Data Presentation:

• Moisture content data is usually presented as a percentage, indicating the proportion of water content in the sample.

5. Applications:

• Food Industry: Moisture content analysis is essential for quality control and shelf-life determination in food products like grains, cereals, fruits, and processed foods.

• Agriculture: It helps in assessing soil moisture levels for irrigation management and crop health.

• Pharmaceuticals: Ensuring the correct moisture content is critical in pharmaceutical manufacturing to maintain drug stability and effectiveness.

• Wood and Paper Industry: It's used to monitor moisture levels in wood products, paper, and cardboard.

- **Construction:** In construction materials like concrete, aggregates, and gypsum, moisture content can affect material properties and curing.

6. Factors Affecting Moisture Content:

- **Environmental Conditions:** Humidity and temperature can influence the moisture content of materials.

- **Material Type:** Different materials have different affinities for moisture absorption or retention.

7. Quality Control:

- In many industries, maintaining the correct moisture content within specified limits is crucial to meet quality standards and regulatory requirements.

8. Regulatory Compliance:

- Some industries, such as pharmaceuticals and food, have strict regulatory requirements for moisture content to ensure product safety and effectiveness.

Moisture content analysis is a fundamental laboratory technique used to assess and control the moisture levels of materials in various industries. Accurate determination of moisture content is critical for product quality, performance, and compliance with regulatory standards.

32

ATTERBERG LIMITS

The Atterberg Limits are a set of standardized laboratory tests used to determine the properties of fine-grained soils, primarily clayey soils. These tests assess the water content at which the soil transitions between different states, providing valuable information for geotechnical engineering, soil classification, and construction projects. The Atterberg Limits consist of three key limits: the Liquid Limit (LL), the Plastic Limit (PL), and the Shrinkage Limit (SL).

1.   Liquid Limit (LL):

•   The Liquid Limit is the water content at which a soil transitions from a liquid-like state to a plastic state, where it begins to behave like a solid and can be molded without crumbling.

•   The test involves using a standardized device called a Casagrande cup. A soil sample is mixed with water, and the cup is repeatedly jarred until the soil exhibits a certain flow behavior.

•   The result is reported as a percentage, indicating the water content at the Liquid Limit.

2.   Plastic Limit (PL):

• The Plastic Limit is the water content at which a soil transitions from a plastic state to a semisolid or semi-liquid state, where it can no longer be molded without breaking or cracking.

• The test involves rolling a soil sample into a thread of a specific diameter to determine the water content at which it crumbles.

• The result is reported as a percentage, indicating the water content at the Plastic Limit.

3. Shrinkage Limit (SL):

• The Shrinkage Limit is the water content at which a soil has the highest water content at which it can shrink when it dries without undergoing further volume changes.

• It is determined by measuring the volume of a soil sample as it undergoes drying and noting the water content at which no further volume changes occur.

• The result is reported as a percentage, indicating the water content at the Shrinkage Limit.

The Atterberg Limits provide important information about the behavior and engineering characteristics of fine-grained soils. They are used for various purposes, including:

- **Soil Classification:** The Atterberg Limits help classify soils into different categories based on their plasticity and compressibility, as defined by standardized classification systems like the Unified Soil Classification System (USCS) and the American Association of State Highway and Transportation Officials (AASHTO) classification system.

- **Construction and Engineering:** Engineers and geotechnical professionals use the Atterberg Limits to assess the suitability of soils for construction projects, including foundations, embankments, and roadways. They also help in determining the optimal moisture content for compaction.

- **Soil Behavior Prediction:** These limits provide insights into how soils will behave under different moisture conditions, helping in predicting issues like settlement, swelling, and shear strength.

In summary, the Atterberg Limits are essential tests in geotechnical engineering and soil science, helping to characterize the properties of fine-grained soils and make informed decisions in construction and land development projects.

33

COMPACTION TESTING

Compaction testing is a geotechnical engineering procedure used to assess the density and moisture content of soil and evaluate its suitability for construction and engineering projects. Compaction is a critical process in the construction of foundations, roadways, embankments, and other structures, as it ensures that the soil has the necessary strength, stability, and resistance to settlement. Here are the key aspects of compaction testing:

1. Purpose:

•	Compaction testing aims to achieve specific soil density and moisture content levels to meet engineering and construction requirements.

•	The primary goals are to increase soil strength, reduce settlement, and improve soil stability.

2. Equipment:

•	Compaction Equipment: Compactors or compacting machines, such as vibratory rollers, sheepsfoot rollers, and plate compactors, are used to compact the soil.

- **Testing Equipment:** Compaction testing requires specialized equipment, including a nuclear density gauge, sand cone test apparatus, or a drive cylinder.

3. Test Methods:

- **Standard Proctor Test:** Also known as the Proctor Compaction Test, it determines the maximum dry density and optimum moisture content of a soil sample under controlled conditions. The test involves compacting soil samples in layers using a standardized compactive effort (usually a standard hammer) and measuring the resulting density and moisture content.

- **Modified Proctor Test:** Similar to the standard Proctor test, but with higher energy compaction, making it more suitable for heavy construction projects.

- **Nuclear Density Test:** This non-destructive test uses a nuclear density gauge to measure soil density and moisture content in the field.

- **Sand Cone Test:** In this test, a hole is excavated in the field, and the volume of the hole is determined using sand of known density. The moisture content and dry density of the soil are calculated based on the volume of the hole and the weight of the soil.

4. Procedure:

• Soil samples are collected from the site and brought to the laboratory for testing.

• In the laboratory, the soil samples are compacted using the specified method and equipment.

• The dry density and moisture content are determined for each compaction effort.

• In the field, nuclear density gauges or sand cone tests are used to assess the in-situ soil compaction.

5. Interpretation:

• The test results provide information on the maximum dry density and optimum moisture content of the soil.

• Engineers use this data to determine the level of compaction needed for the project and assess whether the achieved compaction meets the project specifications.

6. Applications:

• Compaction testing is crucial in the construction of roads, highways, airports, building foundations, embankments, and dams.

• It is used to ensure that the soil meets engineering standards for load-bearing capacity, stability, and settlement control.

7. Quality Control:

• Compaction testing serves as a quality control measure to ensure that the soil used in construction meets the required compaction standards.

8. Regulatory Compliance:

• Many construction projects are subject to regulatory requirements that specify minimum compaction standards to ensure safety and performance.

Proper compaction testing is essential to achieving the desired soil properties for construction and engineering projects. It helps prevent issues like settlement, uneven foundation settling, and structural failure by ensuring that the soil is compacted to the appropriate density and moisture content.

34

SHEAR STRENGTH TESTING

Shear strength testing is a crucial geotechnical engineering procedure used to determine the shear strength properties of soils and rocks. Shear strength is a measure of a material's resistance to shearing forces or stresses, and it is essential for assessing the stability and safety of structures, slopes, foundations, and excavations. Shear strength testing is particularly important in geotechnical and civil engineering projects. Here are the key aspects of shear strength testing:

1. Purpose:

• Shear strength testing aims to determine the material's internal resistance to sliding or shearing under applied forces.

• It provides critical data for designing safe and stable foundations, retaining walls, embankments, and slopes.

2. Equipment and Methods:

• Several laboratory and in-situ tests are used to measure shear strength, including:

- Direct Shear Test: This laboratory test involves applying a shearing force to a soil sample within a shear box apparatus, measuring the horizontal and vertical displacements to calculate the shear strength.

- Triaxial Compression Test: In this laboratory test, a cylindrical soil sample is subjected to radial and axial stresses while controlling pore pressure and measuring deformation. It provides more information about shear strength under different conditions.

- Vane Shear Test: A vane shear device is inserted into the soil, and torque is applied to determine the shear strength of cohesive soils, such as clays.

- Field Vane Test: This in-situ test uses a similar vane device to determine the undrained shear strength of cohesive soils in the field.

- Standard Penetration Test (SPT): Although primarily used for subsurface exploration, the SPT can indirectly provide information about shear strength by assessing soil resistance during penetration.

3. Sample Preparation:

- Soil samples collected from the field are carefully prepared and trimmed to the required size and shape for laboratory shear strength tests.

4. Shear Strength Parameters:

• Shear strength is typically characterized by several parameters, including:

• Cohesion (c): Cohesion is the component of shear strength associated with the cohesive forces between soil particles. It is significant in cohesive soils like clays.

• Internal Friction Angle (φ): The internal friction angle represents the resistance to shearing due to friction between soil particles. It is critical in granular soils like sands and gravels.

• Effective Stress: Shear strength is often discussed in terms of effective stress, which accounts for pore water pressures within the soil.

5. Applications:

• Shear strength testing is used to assess soil stability in various geotechnical engineering applications, including slope stability analysis, foundation design, retaining wall design, and earthwork construction.

6. Safety and Stability:

• Accurate shear strength data is vital for ensuring the safety and stability of structures, excavations, embankments, and natural slopes.

7. Quality Control and Design:

•   Engineers use shear strength data to design foundations and structures that can safely withstand the applied loads and environmental conditions.

8. Regulatory Compliance:

•   Many construction projects are subject to regulatory requirements that specify minimum shear strength values to ensure safety and stability.

Shear strength testing plays a fundamental role in geotechnical engineering by providing critical information about soil and rock behavior under shear forces. It informs the design and construction of civil engineering projects and is essential for ensuring their safety and long-term performance.

35

PERMEABILITY TESTING

Permeability testing, also known as hydraulic conductivity testing, is a geotechnical and hydrogeological procedure used to measure the ability of a soil or porous material to transmit fluids (usually water) through its pores. It is a critical parameter in various fields, including geotechnical engineering, environmental science, hydrogeology, and civil engineering. The results of permeability tests provide valuable information for assessing groundwater flow, designing drainage systems, and evaluating soil characteristics. Here are the key aspects of permeability testing:

1. Purpose:

•   Permeability testing aims to determine how easily fluids can flow through a soil or porous medium under specific conditions.

•   It is essential for understanding groundwater movement, evaluating soil suitability for seepage control, and designing drainage systems.

2. Equipment and Methods:

•   Several laboratory and field tests are used to measure permeability, including:

- **Constant Head Permeability Test:** In this laboratory test, a soil sample is placed in a permeameter, and water is allowed to flow through the sample under a constant head condition. The flow rate and hydraulic gradient are measured to calculate permeability.

- **Variable Head Permeability Test:** Similar to the constant head test, but in this case, the head varies over time as water flows through the sample.

- **Falling Head Permeability Test:** This laboratory test involves measuring the time it takes for the water level to drop in a permeameter column as water flows through the soil sample. It's typically used for less permeable soils.

- **Pump-In Test:** In this field test, water is pumped into a borehole, and the rate at which water level rises is measured to estimate permeability.

- **Slug Test:** A well slug or a volume of water is introduced into a well, and the subsequent water level recovery is monitored to determine permeability.

3. Sample Preparation:

- Soil samples are collected from the field and prepared for laboratory testing. They may be trimmed or compacted to a specific density.

4. Permeability Coefficient (k):

• The permeability coefficient (k) is the primary parameter obtained from permeability testing. It represents the rate of flow of water through the soil and is typically expressed in units like cm/s or ft/day.

5. Soil Properties:

• Permeability is influenced by soil properties such as grain size distribution, porosity, and the presence of clay or other fine-grained materials. Coarser, well-graded soils generally have higher permeability.

6. Applications:

• Permeability testing is used in various applications, including:

• Assessing soil suitability for drainage systems and seepage control in dams and embankments.

• Evaluating the potential for groundwater contamination and flow in environmental site assessments.

• Designing infiltration and drainage systems in civil engineering projects.

7. Regulatory Compliance:

• In environmental and construction projects, regulatory authorities may specify maximum permissible permeability values to prevent groundwater contamination or ensure seepage control.

8. Engineering Design:

•   Engineers use permeability data to design drainage systems, groundwater monitoring programs, and environmental remediation strategies.

Permeability testing is crucial for understanding how water flows through soils and porous materials. It helps in designing effective solutions for managing water movement and ensuring the stability and safety of infrastructure and environmental projects.

36

CONSOLIDATION TESTING

Consolidation testing is a geotechnical laboratory procedure used to determine the settlement characteristics of soils when subjected to applied loads over time. This testing is particularly important in geotechnical engineering and construction projects, as it helps assess the potential for settlement and the time it takes for a soil to undergo consolidation under a given load. Consolidation testing is commonly used in the design of foundations, embankments, and other structures to ensure they remain stable over time. Here are the key aspects of consolidation testing:

1. Purpose:

•   Consolidation testing aims to assess the compressibility and settlement behavior of soils when subjected to vertical loads.

•   It helps in estimating the magnitude and rate of settlement that can occur under applied loads, which is crucial for designing foundations and structures that can accommodate this settlement.

2. Equipment and Methods:

•   Consolidation tests are typically performed using a consolidometer or oedometer apparatus, which consists of a soil sample within a rigid ring, a loading platform, and a mechanism for applying vertical stress.

• The test involves the following steps:

1. A soil sample is prepared and placed within the consolidometer ring.

2. Vertical stress is applied incrementally to the soil sample using dead weights or a hydraulic system.

3. The sample's settlement is measured at specified time intervals as the stress is applied.

4. The resulting data is used to create a consolidation curve, which shows how the settlement changes over time for a given applied stress.

3. Parameters Obtained:

• Consolidation tests provide critical parameters such as:

• Coefficient of Consolidation (C$\square$): A measure of how quickly the soil consolidates under applied loads.

• Coefficient of Volume Compressibility (mv): A measure of how compressible the soil is.

• Preconsolidation Pressure (σp): The maximum past stress that the soil has experienced, beyond which it will undergo settlement without further increase in stress.

•	Compression Index (Cc): A measure of the soil's compressibility.

4. Sample Preparation:

•	Soil samples are collected from the field and carefully prepared in the laboratory to ensure they represent the in-situ conditions.

•	The samples are typically remolded to a specific moisture content and compacted to a standard density.

5. Applications:

•	Consolidation testing is crucial in the design of foundations, embankments, and structures that require long-term stability.

•	It helps predict settlement, which is essential in mitigating potential damage to structures caused by differential settlement.

6. Engineering Design:

•	Engineers use consolidation test data to design foundations that can accommodate anticipated settlement and to assess the suitability of soil for construction.

7. Regulatory Compliance:

A SHORT BRIEF OF EXCAVATION
BY RAMCHANDRA SHRIVASTAV

•   Some construction projects may have regulatory requirements related to settlement limits, which can be determined through consolidation testing.

Consolidation testing provides valuable information about how soils settle over time under applied loads. This information is crucial for designing safe and stable foundations, ensuring the long-term performance of structures, and mitigating potential settlement-related issues in geotechnical engineering and construction projects.

37

CALIFORNIA BEARING RATIO (CBR) TEST

The California Bearing Ratio (CBR) test is a standardized laboratory and field test used in geotechnical engineering to assess the strength of subgrade soils, subbase, and base course materials for road and pavement design. The CBR test provides essential information about the load-bearing capacity of soils and materials, helping engineers design road and pavement structures that can withstand traffic loads. Here are the key aspects of the California Bearing Ratio (CBR) test:

1. Purpose:

• The CBR test aims to determine the relative strength of a soil or material by assessing its resistance to penetration under a standard loading condition.

• It is used for road design, pavement design, and the evaluation of the suitability of subgrade soils and construction materials.

2. Equipment and Test Procedure:

• The CBR test involves the following steps:

A SHORT BRIEF OF EXCAVATION
BY RAMCHANDRA SHRIVASTAV

1.   A cylindrical or square-shaped test mold is prepared and filled with a compacted soil or material specimen.

2.   A standard piston or plunger is then used to apply a specified axial load to the soil specimen at a constant rate.

3.   The penetration depth of the piston or plunger into the specimen is recorded as the load is applied.

4.   The load penetration data is used to calculate the CBR value.

3. CBR Value:

•   The CBR value is expressed as a percentage and is calculated as the ratio of the test load applied to the specimen to the load required to achieve a similar penetration in a standard crushed aggregate material.

•   The standard crushed aggregate material is assigned a CBR value of 100%.

4. Testing Conditions:

•   The CBR test can be performed under various conditions, including dry and soaked conditions.

•   The dry CBR test measures the strength of the soil when it is in its natural moisture state.

• The soaked CBR test measures the strength of the soil after it has been fully soaked to simulate worst-case conditions.

5. Applications:

• The CBR test results are used to design and evaluate flexible and rigid pavements, including roadways, highways, and airport runways.

• Engineers use CBR values to determine pavement thickness, assess subgrade and subbase materials, and estimate pavement performance under traffic loads.

6. Interpretation:

• Higher CBR values indicate greater strength and load-bearing capacity of the soil or material.

• CBR values are typically used in pavement design charts and tables to determine the required pavement thickness for a given traffic load.

7. Regulatory Compliance:

• Many transportation agencies and regulatory bodies specify minimum CBR values for subgrade soils and materials to meet safety and performance standards.

8. Quality Control:

A SHORT BRIEF OF EXCAVATION
BY RAMCHANDRA SHRIVASTAV

• The CBR test is used to ensure that construction materials and subgrade soils meet design specifications and requirements.

The California Bearing Ratio (CBR) test is an essential tool in geotechnical engineering for pavement design and quality control in road and highway construction. It helps engineers assess the load-bearing capacity of soils and materials, ensuring the safe and reliable performance of transportation infrastructure.

38

SWELL POTENTIAL TESTING

Swell potential testing, also known as soil swelling testing, is a geotechnical procedure used to assess the potential for soil or clayey materials to expand and contract when subjected to changes in moisture content. Swelling soils can pose significant challenges in construction and civil engineering projects, as they can lead to ground movement, foundation damage, and infrastructure problems. Swell potential testing helps engineers and geotechnical professionals evaluate the behavior of these soils and make informed decisions regarding construction and site development. Here are the key aspects of swell potential testing:

1. Purpose:

•   Swell potential testing aims to determine the extent to which a soil or clayey material may expand when it absorbs water and contract when it dries out.

•   It is used to identify soils that may exhibit significant volume changes due to moisture variations, which can be problematic for construction projects.

2. Equipment and Test Methods:

•	The test methods for assessing swell potential can vary, but common laboratory tests include:

•	Free Swell Test: In this test, a soil sample is compacted to a specific density and moisture content and then immersed in water. The volume change that occurs as the soil swells is measured.

•	Shrinkage Limit Test: The shrinkage limit is determined by measuring the water content at which the soil begins to shrink upon drying. Soils with a higher shrinkage limit are more prone to swelling.

•	Atterberg Limits: The Atterberg Limits, specifically the Liquid Limit and Plastic Limit, can provide information about the soil's potential for volume change.

3. Swell Potential Classification:

•	Swell potential is often classified into different categories based on the extent of soil expansion:

•	Low Swell: Minimal volume change with changes in moisture content.

•	Moderate Swell: Moderate volume change but manageable with proper engineering measures.

• **High Swell:** Significant volume change that can be challenging to mitigate.

4. Applications:

• Swell potential testing is primarily used in construction projects to assess soil suitability for foundations, pavements, embankments, and other structures.

• It helps engineers identify and address potential swelling issues before construction begins.

5. Engineering Design:

• Engineers use swell potential data to design foundations and structures that can accommodate soil swelling or to implement soil modification techniques to mitigate swelling.

6. Mitigation Measures:

• For soils with high swell potential, engineers may implement mitigation measures, such as moisture control, chemical stabilization, or the use of geosynthetics, to reduce or prevent swelling.

7. Regulatory Compliance:

• Regulatory authorities may specify swell potential testing and mitigation measures for construction projects to ensure safety and stability.

A SHORT BRIEF OF EXCAVATION
BY RAMCHANDRA SHRIVASTAV

Swell potential testing is essential for identifying and managing the challenges posed by swelling soils in construction and civil engineering projects. By understanding the behavior of these soils, engineers can develop appropriate strategies to mitigate potential issues and ensure the long-term performance and safety of infrastructure.

39

CHEMICAL ANALYSIS

Chemical analysis refers to the process of determining the composition and properties of substances and materials by examining their chemical and physical characteristics. This analytical technique is widely used in various scientific fields, industries, and research areas to understand, identify, and quantify the chemical components of substances. Chemical analysis can provide valuable information about the purity, quality, and behavior of materials. Here are some key aspects of chemical analysis:

1. Purpose:

•   Chemical analysis serves a wide range of purposes, including:

•   Quality Control: Ensuring products meet specific quality standards and regulatory requirements.

•   Research and Development: Developing new materials, formulations, and processes.

•   Environmental Monitoring: Assessing pollutant levels, monitoring air and water quality, and evaluating soil composition.

- **Food and Pharmaceuticals:** Ensuring the safety and quality of food products and pharmaceuticals.

- **Forensic Science:** Identifying substances in criminal investigations.

- **Geological and Environmental Studies:** Analyzing rock and soil samples for mineral composition and pollution assessment.

2. Analytical Techniques:

- **Various analytical techniques** are employed for chemical analysis, depending on the specific application and the properties of the material. Some common techniques include:

- **Spectroscopy:** This includes techniques like atomic absorption spectroscopy (AAS), ultraviolet-visible spectroscopy (UV-Vis), and infrared spectroscopy (IR).

- **Chromatography:** Gas chromatography (GC), liquid chromatography (LC), and high-performance liquid chromatography (HPLC) are used to separate and quantify chemical compounds.

- **Mass Spectrometry (MS):** MS is used to identify and quantify compounds based on their mass-to-charge ratio.

• **Titration: A technique used to determine the concentration of a solution by reacting it with a solution of known concentration.**

• **Electrochemical Analysis: Techniques like potentiometry and voltammetry are used to measure electrochemical properties.**

• **Gravimetry: Analyzing substances based on their mass, often through precipitation and filtration.**

3. Sample Preparation:

• **Proper sample preparation is crucial for accurate chemical analysis. It may involve:**

• **Homogenizing samples to ensure they are representative.**

• **Extracting compounds of interest.**

• **Diluting samples to suitable concentrations.**

4. Data Interpretation:

• **The data obtained from chemical analysis is processed and interpreted to determine the composition, concentration, and other relevant properties of the substances being analyzed.**

• **Calibration and standardization are often necessary to ensure accuracy.**

5. Applications:

• Chemical analysis is applied in a wide range of fields, including chemistry, biology, environmental science, materials science, food science, pharmaceuticals, and more.

• It plays a critical role in quality control, research, and regulatory compliance.

6. Instrumentation:

• Modern chemical analysis relies heavily on advanced instrumentation and equipment, including spectrometers, chromatographs, mass spectrometers, and various detectors.

7. Regulatory Compliance:

• Many industries are subject to regulatory requirements that mandate specific chemical analyses to ensure safety, quality, and compliance with standards.

Chemical analysis is a fundamental tool in science and industry, providing critical information about the composition, quality, and behavior of materials. It enables researchers and professionals to make informed decisions and advancements across various disciplines.

40

STANDARD PENETRATION TEST (SPT)

The Standard Penetration Test (SPT) is a widely used in-situ geotechnical testing method for assessing the subsurface properties of soils and determining their engineering characteristics. It is an essential tool in geotechnical engineering and is commonly performed during site investigations for construction, foundation design, and infrastructure projects. The SPT provides valuable information about soil stratigraphy, relative density, and shear strength. Here are the key aspects of the Standard Penetration Test (SPT):

1. Purpose:

•	The primary purpose of the SPT is to collect information about the engineering properties of soils at various depths below the ground surface.

•	It helps in characterizing soil layers, assessing soil strength, and designing foundations and other structures.

2. Equipment and Procedure:

•	The SPT involves the following equipment and procedure:

1.   A borehole is drilled to the desired depth using drilling equipment such as an auger or rotary drill.

2.   A split-spoon sampler (SPT sampler) is driven into the soil at the bottom of the borehole using a hammer with a standard weight and fall height.

3.   The number of hammer blows required to drive the sampler a standard distance (usually 12 inches or 300 mm) into the soil is recorded as the "N-value."

3. N-Value:

•   The N-value represents the number of blows required to achieve a standard penetration distance and is a measure of soil resistance to penetration.

•   It is used as an indicator of relative soil density and to classify soil types according to recognized standards.

4. Soil Classification:

•   The SPT N-value is used to classify soil types according to systems such as the Unified Soil Classification System (USCS) and the American Association of State Highway and Transportation Officials (AASHTO) classification system.

5. Applications:

- The SPT is used in various geotechnical applications, including:

- Foundation design: Determining bearing capacity and settlement estimates.

- Slope stability analysis: Assessing soil strength for embankments and slopes.

- Liquefaction assessment: Evaluating the potential for soil liquefaction during seismic events.

- Ground improvement: Designing ground improvement techniques for problematic soils.

6. Data Interpretation:

- The N-values obtained from the SPT are analyzed to interpret soil properties, estimate soil shear strength, and evaluate geotechnical parameters such as cohesion and friction angle.

- The data can also be used to construct soil profiles and assess soil behavior under different loading conditions.

7. Safety Considerations:

- Safety precautions must be taken during SPT testing to ensure the well-being of field personnel, especially when using heavy equipment and operating in unstable or loose soils.

8. Limitations:

•   The SPT provides valuable information but has limitations, such as not directly measuring parameters like shear strength. Additional laboratory testing may be needed for more accurate assessments.

The Standard Penetration Test (SPT) is a widely accepted and valuable tool in geotechnical engineering. It provides engineers and geotechnical professionals with critical information about subsurface soil properties, which is essential for designing safe and stable foundations and infrastructure projects.

41

CONE PENETRATION TEST (CPT)

The Cone Penetration Test (CPT) is a versatile and widely used in-situ geotechnical testing method for characterizing the subsurface soil and sediment properties. It involves driving a cone-shaped probe (the penetrometer) into the ground at a constant rate while continuously measuring the resistance to penetration and pore water pressure. The CPT provides valuable information about soil stratigraphy, soil properties, and geotechnical parameters, making it a valuable tool in geotechnical engineering and environmental investigations. Here are the key aspects of the Cone Penetration Test (CPT):

1. Purpose:

• The primary purpose of the CPT is to assess the geotechnical and geophysical properties of soils and sediments at various depths below the ground surface.

• It is used for site investigations, foundation design, soil classification, and environmental assessments.

2. Equipment and Procedure:

A SHORT BRIEF OF EXCAVATION
BY RAMCHANDRA SHRIVASTAV

• The CPT involves the following equipment and procedure:

1. A penetrometer, typically a cone-shaped tip attached to a rod or string, is advanced into the soil at a constant rate using hydraulic or mechanical equipment.

2. As the penetrometer advances, it continuously measures the cone's resistance to penetration (cone tip resistance or q_c) and the local pore water pressure (u) using sensors.

3. The data obtained during the test is transmitted to the surface and recorded for analysis.

3. Parameters Obtained:

• The CPT provides several key parameters, including:

• Cone Tip Resistance (q_c): This is a measure of the soil's resistance to penetration and can be related to soil strength and bearing capacity.

• Friction Sleeve Resistance (f_s): The resistance experienced by the friction sleeve on the penetrometer, which provides additional information about soil properties.

• Pore Pressure (u): The measurement of pore water pressure helps assess soil consolidation and provides insights into soil compressibility.

- **Sleeve Friction Ratio (R_f):** The ratio of friction sleeve resistance to cone tip resistance, which can help identify soil types.

- **Sleeve Friction (f_s):** The total friction resistance experienced by the penetrometer.

4. Soil Classification:

- The CPT data is used to classify soil types and assess soil stratigraphy based on recognized standards and correlations with soil properties.

5. Applications:

- The CPT has a wide range of applications, including:

- **Foundation design:** Evaluating bearing capacity and settlement predictions.

- **Geotechnical investigation:** Assessing soil behavior for embankments, slopes, and retaining walls.

- **Environmental assessments:** Identifying contamination, monitoring groundwater, and evaluating soil properties for environmental purposes.

6. Data Interpretation:

- The CPT data is analyzed to interpret soil properties, estimate geotechnical parameters, assess soil behavior under different loading conditions, and make engineering decisions.

7. Safety Considerations:

•   Safety precautions must be taken during CPT testing to protect field personnel and ensure the proper operation of equipment.

8. Advantages:

•   The CPT offers several advantages, including continuous data collection, real-time results, and minimal soil disturbance compared to other in-situ testing methods.

The Cone Penetration Test (CPT) is a versatile and valuable tool in geotechnical engineering and environmental investigations. Its ability to provide continuous, real-time data makes it particularly useful for assessing subsurface conditions and making informed decisions in construction, foundation design, and environmental assessments.

42

VANE SHEAR TEST

The Vane Shear Test, also known as the Vane Shear Strength Test, is a geotechnical laboratory and in-situ test used to determine the shear strength properties of cohesive (clayey) soils. It specifically measures the undrained shear strength of these soils, which is essential for assessing their stability and load-bearing capacity. The test involves rotating a vane-shaped blade within a soil sample and measuring the torque required for shearing. Here are the key aspects of the Vane Shear Test:

1. Purpose:

•   The Vane Shear Test is primarily used to assess the undrained shear strength of cohesive soils, particularly clays.

•   It helps geotechnical engineers and researchers evaluate the stability of slopes, embankments, foundations, and retaining walls built on or with these soils.

2. Equipment and Procedure:

•   The test involves the following equipment and procedure:

A SHORT BRIEF OF EXCAVATION
BY RAMCHANDRA SHRIVASTAV

1.   A cylindrical soil sample is collected from the field or prepared in the laboratory. The sample is typically undisturbed to maintain its natural properties.

2.   A vane apparatus, consisting of a vane blade attached to a rod, is inserted into the soil sample.

3.   The vane blade is then rotated at a constant rate (usually 6 degrees per minute) while the torque required to shear the soil is continuously measured.

4.   The test continues until failure occurs, which is indicated by a rapid increase in torque.

5.   The maximum torque recorded during the test is used to calculate the undrained shear strength of the soil.

3. Undrained Shear Strength:

•   The undrained shear strength (often denoted as cu) is a measure of a soil's resistance to shearing without undergoing significant drainage or consolidation.

•   It is a critical parameter for assessing the stability of cohesive soils, especially during rapid loading events, such as earthquakes or construction activities.

4. Sample Preparation:

- For accurate testing, the soil sample must be carefully prepared to maintain its in-situ structure and moisture content.

5. Applications:

- The Vane Shear Test is used in various geotechnical applications, including:

- Foundation design: Assessing soil stability for shallow foundations.

- Slope stability analysis: Evaluating the stability of slopes and embankments.

- Retaining wall design: Determining soil strength for retaining wall stability.

- Seismic analysis: Assessing soil behavior during earthquake events.

6. Interpretation:

- The undrained shear strength obtained from the Vane Shear Test is compared to design requirements to assess the suitability of the soil for specific engineering purposes.

- It is used to estimate the safety factor for various geotechnical designs.

7. Safety Considerations:

A SHORT BRIEF OF EXCAVATION
BY RAMCHANDRA SHRIVASTAV

• Safety precautions are essential when operating vane shear equipment to avoid injury to personnel and ensure accurate testing.

The Vane Shear Test is a valuable tool for assessing the undrained shear strength of cohesive soils, particularly clays. By providing data on soil strength, it helps geotechnical engineers make informed decisions regarding the stability and safety of structures and embankments built on these soils.

43

PLATE LOAD TEST

The Plate Load Test is a field test used to determine the bearing capacity and settlement characteristics of soils beneath foundations, pavements, and other loaded structures. It is a crucial geotechnical test that helps engineers assess the soil's ability to support loads and provides valuable data for foundation design and construction. Here are the key aspects of the Plate Load Test:

1. Purpose:

•　　The primary purpose of the Plate Load Test is to assess the load-bearing capacity of the soil and predict the settlement that will occur under a specified load.

•　　It helps engineers design foundations and pavements that can safely distribute loads without excessive settlement or deformation.

2. Equipment and Procedure:

•　　The Plate Load Test involves the following equipment and procedure:

1.　　A steel plate, often called the "loading plate" or "bearing plate," is placed on the ground surface where the foundation or pavement will be constructed.

2.   The plate is loaded with a known vertical force or load, typically applied incrementally using hydraulic jacks.

3.   The load is increased until the desired load levels are reached or until the soil undergoes excessive settlement or failure.

4.   During the test, measurements are taken to monitor the settlement of the plate, as well as the corresponding load applied.

5.   The data collected during the test is used to plot a load-settlement curve, which provides valuable information about the soil's behavior under load.

3. Parameters Obtained:

•   The Plate Load Test provides several important parameters, including:

•   Ultimate Bearing Capacity (q): This is the maximum load that the soil can support without excessive settlement or failure.

•   Settlement: The settlement of the plate under the applied load is measured at various load levels.

•   **Modulus of Subgrade Reaction (k): The stiffness of the soil beneath the plate can be calculated from the load-settlement data.**

4. Applications:

•   **The Plate Load Test is used for various geotechnical and civil engineering applications, including:**

•   **Foundation design: Assessing the soil's capacity to support building foundations.**

•   **Pavement design: Evaluating the suitability of subgrade soils for road and pavement construction.**

•   **Soil improvement: Determining the effectiveness of ground improvement techniques.**

•   **Retaining wall design: Assessing the stability of retaining wall foundations.**

5. Interpretation:

•   **Engineers analyze the data from the Plate Load Test to determine the appropriate foundation type, size, and depth, as well as the expected settlement of the structure.**

•   **The results are used to ensure the safety and stability of the designed structure under the anticipated loads.**

6. Safety Considerations:

• Safety precautions are essential when conducting Plate Load Tests to prevent accidents during loading and to ensure the safety of field personnel.

The Plate Load Test is a valuable field test in geotechnical engineering for assessing the bearing capacity and settlement behavior of soils under various loading conditions. It plays a critical role in the safe and economical design of foundations, pavements, and other civil engineering structures.

THE END

A SHORT BRIEF OF EXCAVATION
BY RAMCHANDRA SHRIVASTAV